ROCKET AND SPACE CORPORATION ENERGIA

The Legacy of S. P. Korolev

All rights reserved under article two of the Berne Copyright Convention (1971).
We acknowledge the financial support of the Government of Canada through the
Book Publishing Industry Development Program for our publishing activities.
Published by Apogee Books an imprint of Collector's Guide Publishing Inc., Box 62034, Burlington, Ontario, Canada, L7R 4K2
Printed and bound in Canada
Rocket & Space Corporation Energia
Edited by Robert Godwin
ISBN 1-896522-81-5
English Edition ©2001 RSCE/Apogee Books
All photos courtesy of RSCE

FROM FIRST SATELLITE TO ENERGIA - BURAN and MIR

Editor's Introduction

When I first acquired an imprint of the original Russian edition of this book I sat and perused the pages in stunned silence. I felt as though I had fallen into some kind of strange parallel universe. Within the pages were pictures of things familiar and yet not so.

It is perhaps a testament to the ingenious human spirit that two entirely divergent cultures could make such remarkable strides in the field of space exploration and yet indelibly stamp their own mark on the designs. The inexorable laws of physics dictate that there are certain absolutes which constrain us, but the fact remains that there are many ways to achieve the same goals.

In the following pages you will see images which bring to life the dextrous faculties of the Russian engineers and scientists. Arrayed within are an impressive string of designs which placed the Russian people in the vanguard of space exploring nations.

In much the same way as the United States had done, the victorious forces of the Soviet Union returned home at the end of World War 2 with the spoils of war. Accessing much of the remarkable research undertaken by the German scientists at Peenemünde the great designer S.P. Korolev brought the R-1 missile to life and placed his country on a road peppered with historic accomplishments.

From 1957's first artificial satellite Sputnik through to the remarkable space stations of the end of the 20th century the Russian people and the engineers and scientists of Rocket & Space Corporation Energia have created and sustained an impressive legacy of technological triumphs.

At the turn of the millennium the Russian and American people are now working together with people from around the world to establish the International Space Station. Undoubtedly this synergy between East and West has only just begun to bear fruit and the world has yet to see where this new détente between old adversaries will lead us. One thing is clear however, the aptitude and excellence of the designs which continue to emerge from Rocket and Space Corporation Energia will continue to surprise us.

Robert Godwin
(Editor - English Edition)

Special thanks for assistance with the English edition to:

Space Media Incorporated and
Space Hab Incorporated

Two significant events laid the foundation for the epoch of space exploration: launch into Earth orbit of the world's first artificial satellite (October 4, 1957); and the first manned Earth-orbiting space flight (April 12, 1961). With these landmark events, the evolution of national cosmonautics entered the history of mankind.

Preceding these events, much hard work was undertaken in the development of rocket and space technology, and its associated industries, beginning as early as 1946.

In the spring of 1946, NII (Research Institutes), KB (Design Bureaus), and test centers were created in accordance with a government decision, and plants for the development and manufacture of ballistic long-range missiles were conceived.

88 State Research Institute of Reaction Armament (NII-88) (which in 1956 became the OKB-1 independent organization and now is called S. P. Korolev space corporation Energia) acted as the prime organization for this work. At that time, a team led by General Designer Sergei Pavlovich Korolev was engaged in the design of ballistic long-range missiles with liquid rocket engines.

While complying with state assignments to create combat long-range missiles, S. P. Korolev oriented his team to simultaneously develop and perform space exploration study programs beginning with research of the Earth's upper atmospheric layers. Therefore, after the flight of the first native ballistic missile, R-1 (October 10, 1948), flights of R-1A, R-1B, R-1V and other

geophysical rockets followed. After the successful launch of the world's first intercontinental ballistic missile, R-7 (August 21, 1957), launches of the first Earth artificial satellites were performed, as well as launches of spacecraft of various purpose using modified R-7 missiles. Wide-scale exploration of space had begun: Luna, Venera, Mars, Zond and other automatic interplanetary stations were launched; flights of unmanned and manned spacecraft of the Vostok type were made; multi-seat spacecraft of the Voskhod type were created; and the first cosmonaut egress into open space was carried out.

As the research scope was widened and studies were extended, Korolev delegated specific research and development subjects to other organizations, transferring to them his deputies and the best qualified personnel to continue the work begun. For example, all matters related to communication satellites he referred to the KB led by M. Ph. Reshetnev; subjects of probing and photography of the Earth to D. I. Kozlov; problems caused by studies of deep space and automatic Earth artificial satellites to G. N. Babakin; and so on, keeping manned spacecraft and heavy launch vehicles for himself. Therefore, practically all of the KB's engaged in the field of space technology originated with, but were then separated from, the KB led by Korolev himself.

The team of S. P. Korolev, continuing his traditions, created a new series – the Soyuz spacecraft – with which the docking of spacecraft in orbit was tested, allowing crew members to transfer from one spacecraft to another.

At the beginning of the 1970's NPO Energia (the former Korolev KB) was headed by academician V. P. Glushko.

At this time a new stage of orbital station creation was begun. The problems involved in long-term station operation were solved. Crew rotation and cargo delivery were performed using both manned and cargo spacecraft.

The Mir station, to which the Kvant, Kvant-2, and Kristall research modules were later docked, was in orbit from February 20, 1986 until its successful deorbit in 2001. The work performed at orbital stations provided great scientific and national economic value. International crews took part in flights to the orbital stations.

The Energia launch vehicle, combined with rocket boosters created at NPO Energia, allowed a universal space platform, inside a cargo transport container, to be put into near-Earth orbit to solve several tasks of national economic purpose, including the creation of global communication system, Telecast. It also put automatic interplanetary spacecraft into flight trajectories to the Moon, the planets and deep space, providing both new, powerful acquisitions of scientific knowledge and practical human activity in the study and exploration of space.

The need for reducing the cost of injecting payload mass into orbit is the main stimulating factor for further modifications and creation of new launch vehicles. Zenit and Energia-M launch vehicles, developed on the basis of the Energia system, allow this task to be solved.

The national space program has always envisaged cosmonautics as being used not only in the interests of our country, but in those of all mankind.

RSCE stands ready to exchange its achievements in space with all countries. We propose performing launches of spacecraft of other nations and international organizations with our launch vehicles and carrying out joint studies at orbital stations, based on mutual agreement.

As always, we shall do everything to keep space peaceful, international, and serving the interests of all mankind, both now and in the future.

President of S. P. Korolev Space
Corporation Energia
Yu. P. Semenov

Space for science, only for peaceful purposes, for the benefit of a man relentlessly perceiving the innermost mysteries of nature – that is the way space studies are developed and performed.

S. P. Korolev

In 1946 S. P. Korolev was charged with heading the development work on ballistic liquid-propellant long-range missiles.

Having gained experience with the prototype research missiles of the pre-war period and having studied the problems with the German missile weapons, Korolev began his own independent path of development. He created a number of native teams within the rocket-space complex, heading up the manned spacecraft and heavy launch vehicle development group himself.

To provide operational solutions to all of the various fundamental scientific and technical problems encountered in the course of developing the missile complexes, Korolev initiated the Council of Chief Designers, including S. P. Korolev, V. P. Barmin, V. P. Glushko, V. I. Kuznetsov, N. A. Piljugin, and M. S. Rjazansky. Each Chief Designer headed his own KB (Design Bureau), each with a different specialty.

The first controlled ballistic long-range missile, the R-1, was developed by the Korolev team based on the German A-4 (V-2) rocket in 1948.

The R-1 missile was 13.4 tons in mass, had a 270 km range, and a non-separating nose cone with a mass of 1.1 tons. The R-1 missile engine, RD-100, was created based on the German rocket engine at the Glushko KB. Liquid oxygen and alcohol were used as the propellant. Missile flight control was performed using aerodynamic vanes and gas control jets.

13 NII's (Research Institutes) and KB's, as well as 35 plants, took part in the creation of the R-1 missile. The first launch of the R-1 occurred on September 17, 1948. It failed. Because of a control system failure the missile deviated almost 50° from the flight line. Success came with an October 10, 1948 launch. In 1950, after completion of flight design tests, the R-1 missile was put into operation with its ground support complex.

Hand-in-hand with the creation of combat ballistic missiles, on Korolev's initiative, a program to research the upper atmosphere was developed in partnership with institutes of the Academy of Sciences of the USSR. Based on the R-1 missile, R-1A, R-1B, R-1V, R-1E and other geophysical rockets were created. Using these missiles comprehensive studies of the atmosphere up to an altitude of 100 km were carried out. On April 21, 1949 the first geophysical rocket, the R-1A, lifted two containers with scientific equipment to an altitude of 110 km, they were then recovered using parachutes.

Further work on ballistic missiles led to the R-2 in 1950. To increase the accuracy, the missile nose cone, 1.5 tons in mass, was made separable during flight. The R-2 range was 590 km with a launching mass of 20.3 tons. Thus, in 1951, a second missile complex was put into operation for the Soviet Army.

Based on the R-2 missile, the R-2A geophysical rocket was created which performed atmospheric probing up to an altitude of 210 km.

In 1953 the first tactical missile using a storable propellant (nitric acid and carbon-hydrogen fuel), the R-11, was created with a range of 270 km. The R-11's launch mass was 5.5 tons and the nose cone mass was 0.67 tons. The engine thrust was about 8 tons with the system propellant developed by the Isaev KB installed on the missile. The thrust vector control was performed by gas jet. The first launch of the R-11 missile occurred on April 18, 1953. In 1955 the missile was put into operation.

The R-11 was the basis of development of the R-11M and R-11FM missiles. The R-11M missile was designed to use a nose cone with a military nuclear charge. The first launch of the R-11M missile was performed on December 30, 1955. A complex with R-11M missiles was put into operation in 1958.

The R-11FM missile was designed to be launched from submarines. The R-11FM was first launched from the swinging sea stand in May 1955, and then on September 16, 1955 from a submarine. The missile was launched from submarine above the water line. The R-11FM missile opened up a new trend of development in combat missiles – sea-based missiles – and was put into operation by the USSR Navy. Further work on sea-based missiles was transferred to a newly organized KB which was headed by V. P. Makeev, successor of S. P. Korolev. The missile was first launched from underwater on December 23, 1958.

Work on creation of ballistic long-range missiles continued at OKB-1 and, as a result of goal-oriented studies and experiments, the first strategic missile, the R-5, appeared. The first R-5 launch was on March 15, 1953, with a range of 1200 km. A liquid oxygen and alcohol engine of 43.8 tons thrust at ground level was installed on the missile. Flight control was performed by gas jets and aerodynamic surfaces.

In 1955, a modification of this missile, the R-5M, with a nuclear military charge in the nose cone, was developed. The first launch of the R-5M missile was on January 21, 1955 and its testing with a nuclear military charge was carried out on February 2, 1956. In 1956 the R-5M missile was put into operation.

Along with the R-5 and R-5M missiles, geophysical rockets R-5A, R-5B, R-5V, and R-5R were created and used to continue studies of the upper atmospheric layers and space, as well as to investigate advanced rocket performance. On February 21, 1958 the R-5V rocket lifted scientific equipment with a mass of 1520 kg to an altitude of 473 km – a record at the time.

The creation of the two-stage intercontinental ballistic missile, the R-7, was the outstanding achievement of native rocket development. The launch mass of the R-7 missile was 280 tons. Unlike preceding missiles, the launching facilities for the R-7 missile were stationary. Launch of this missile from USSR territory could respond to nuclear attack from practically any point in possible enemy territory.

The first stage of the R-7 consisted of four side units. The second stage core unit also included an upper compartment in which a payload of up to 5.4 tons was accommodated. The main four-chamber engines designed by V. P. Glushko and new control engines from S. P. Korolev for controlling the thrust vector were installed in these units. All engines used liquid oxygen and kerosene. The drive for the turbopump units was actuated using hydrogen peroxide. The engines of all units were started on the ground. The liftoff thrust was 406 tons.

Because of its overall dimensions, the missile was delivered to the testing grounds by rail in a disassembled state. The missile assembly, with further pneumo- and electro-tests, was carried out at the technical complex specially built for this purpose. The assembled and tested missile was transported to the launching site by railway line using a special transport-installation unit propelled by a diesel locomotive. The loading of the missile propellant components was carried out from mobile loading units delivered to the launching site after the missile.

The first launch of the R-7 missile, on May 15, 1957, was a failure. The R-7 successfully flew to intercontinental range on August 21, 1957. There was a special TASS report on this launch, which was the third after the flight tests began, informing the world that the Soviet Union had become the owner of this lethal weapon.

In January 1956, on S. P. Korolev's insistence, a decision was made to develop an artificial Earth satellite which could be launched by the R-7 missile. The fact of this launch was to be communicated to and verified by all of the countries of the world. For this purpose, radio equipment was installed on the satellite. Accurate measurement of the orbit parameters of the artificial satellite was provided by radio and optical stations.

The world's first artificial orbiting satellite was injected into near-Earth orbit by an R-7 launch vehicle on October 4, 1957. This event marked the beginning of a new era in the history of civilization – the space age.

Earth's first artificial satellite (PS-1, 83.6 kg in mass) went into an orbit with an apogee of 947 km, a perigee of 228 km, an inclination of 65.6°, and remained in orbit for 92 days. This first Earth orbiting satellite provided data on the lifetime of satellites in near-Earth orbit, on radio wave passage through the ionosphere, and on the effects of space flight conditions on satellite equipment operation. A month later (November 3, 1957) the second Earth artificial satellite (PS-2, 508.3 kg in mass) was put into orbit with an experimental animal (a dog, Laika) on board, and then on May 15, 1958 the third Earth satellite (D-1, 1,327 kg in mass – a real space laboratory) was launched into space.

The results of these first Earth satellite launches were the genesis of the development of interplanetary stations to investigate the Moon and planets of the Solar System.

Missions for study of the Moon and interplanetary flight required re-equipping the launch vehicle with a third stage to increase its power-mass characteristics. In addition to the third stage a booster, which could impart an additional cosmic speed (more than 11 km/s) to interplanetary stations, was required to enable them to depart from Earth orbit.

Interplanetary stations (IS's) and automatic interplanetary stations (AIS's) were designed for flights to the Moon.

They were given the drawing symbol "E":
- IS E1 – for Moon flyby;
- IS E1A – For reaching the surface of the Moon;
- AIS E2, E2A, E3 – for Moon fly-around and photography of its back side;
- AIS E6 – for soft landing on the Moon's surface with transfer of its surface images to Earth;
- AIS E7 – for creation of a Moon satellite;
- AIS E8 – for provision of soft landing on the Moon, and soil sampling and its delivery to Earth.

Stations E1, E2, and E3 were to be launched by a three-stage rocket (R-7 plus rocket unit E) and E6 and the following by a four-stage rocket (R-7 plus rocket unit I plus booster L).

The first native liquid-fueled rocket engine used a liquid oxygen and carbon-hydrogen propellant, provided 5.6 tons of thrust, and was installed in unit E. To drive the turbopump unit, gas pressure was derived from a generator that used the main propellant components. A system of special gas distribution throttles, gas lines and control gas-reaction nozzles behind the turbine was first used for control on unit E. The engine development was jointly conducted by the S. P. Korolev and S. A. Kosberg KB's.

Unit I was also used as the third stage of a four-stage launch vehicle and designed for the spacecraft's final maneuvering into Earth satellite orbit. A four-chamber liquid oxygen and carbon-hydrogen propellant engine, the RO-9 providing 30 tons of thrust, was installed in the unit. This engine was developed by the Kosberg KB.

Booster L was conceived for boosting a spacecraft out of Earth orbit and transferring it into a planetary flight trajectory. For the first time, a rocket unit was fired under weightlessness. The world's first closed-loop engine, with thrust of about 7 tons and using liquid oxygen and carbon-hydrogen propellant, was installed in booster L. This engine was developed by the Korolev KB.

IS's of the E1 and E1A types differed mainly in the scientific equipment installed. Structurally they were similar to the first Earth satellite PS-1.

AIS's E2, E2A, and E3 had solar array elements, radio complex antennas, and gas microengines for altitude control on the outer surface. The radio complex, automation, research equipment, phototelevision device, and buffer electric batteries were housed inside the main hull.

AIS E6 differed from its predecessors. It consisted of three main, functionally isolated parts:

- a correcting-brake engine with control system units;
- two compartments with equipment that were jettisoned before braking at the Moon's surface;
- an automatic autonomous lunar station.

None of the systems of AIS E6 were duplicated because of strict mass limitations.

The first successful launch of an IS – E1, known in the press as "Mechta" (Luna-1) – was performed on January 2, 1959. This station flew at a distance of 5 to 6 thousand km from the Moon and then became a satellite of the Sun. IS E1A started on September 12, 1959 and delivered a pennant of the Soviet Union to the Moon on September 14, 1959. This station was named Luna-2. Luna-3, launched on October 4, 1959, spent 40 minutes photographing the back side of the Moon and then transferred its imagery to Earth. The world's first television image of the Moon's surface was obtained by AIS Luna-9, launched from Earth on January 31, 1966.

Automatic stations of the type 1M (to Mars), 1VA (to Venus), and then MV, the launch of which was performed by the above mentioned four-stage rocket (R-7 plus unit I plus booster L) were designed for flights to Mars and Venus. Activity on creation of these stations began in 1960.

The first four-stage rocket and space system with the 1M-type automatic interplanetary station (AIS) aboard for exploring Mars was launched on October 10, 1960. Because the I rocket module engine failed, the AIS was not injected into Earth orbit. On February 12, 1961 the 1VA-type AIS

was launched to study Venus and flew to within a distance of about 100 thousand km from the planet. This AIS was named Venera-1.

Because of the tasks identified for solution with respect to exploration of interplanetary space – planet fly-by's, with photography and radio probing at small distances, and delivery of the descent vehicles to the planet's surfaces – it was decided to proceed to the development of the MV-type unified automatic interplanetary station for flights to Mars and Venus.

On November 1, 1962 an MV-type station (2MV-4 No 4) named Mars-1 with a mass of 893.5 kg was launched. However, because of deficient pressurization of the high pressure system for operating the altitude-control microengines the station failed to fulfill its task. All subsequent 2MV-type AIS's were not successful either.

AIS 3MV-4 No 3 (Zond-3), launched into heliocentric orbit with a Moon fly-by on July 18, 1965, was the first AIS to completely fulfill its task. On November 12, 1965 the Venera-2 AIS was launched into Venus fly-by trajectory, and on March 1, 1966 the Venera-3 AIS (3MV-3 No 1, launched on November 16, 1965), delivered a Soviet Union pennant to the surface of Venus.

The successful missions of Zond-3, Venera-2, and Venera-3 made it possible to terminate the first phase of the planned program of Mars and Venus exploration and draw a number of fundamental scientific conclusions, specifically: to determine the boundary of Earth's atmosphere; to clarify the character of magnetic fields in the Solar System; and to give the first insights into the atmospheres of the planets explored.

In 1966, all work related to the exploration of the Solar System's planets and the Moon using automatic interplanetary stations (including continuation of work on E6, E7 and E8-type AIS's) was transferred to the KB headed by G. N. Babakin.

The daring idea was carried further as

preparation for the first manned spacecraft launch began. In the spring of 1957 in OKB-1 (as Korolev KB came to be called) the spacecraft design department under the supervision of M. K. Tikhonravov was organized for the purpose of studying and deriving solutions for the complex problems relating to launching a man into space. Having conducted extensive studies since September 1958, this department started passing the technical directions on the development of the spacecraft's onboard systems to its co-executors. Tedious work on the development and testing of the spacecraft, rocket, and launching complex systems was culminated by check launches of the 1 KP unmanned spacecraft (May 15, 1960) and spacecraft with dogs aboard (Chaika and Lisichka on July 28, 1960, Belka and Strelka on August 19, 1960, Pchelka and Mushka on December 1, 1960, Shutka and Cometa on December 22, 1960, Chernushka on March 9, 1961, and Zvezdochka on March 25, 1961) and using dummies.

The test flights were not without problems. For various reasons, the program was twice interrupted (on July 28, 1960 and December 1, 1960), and the flight of the spacecraft-satellite launched on December 22, 1960 became only a suborbital mission. The causes of the failures were thoroughly analyzed and eliminated.

The experience gained made it possible to proceed immediately to preparation for launching a manned spacecraft. The Vostok spacecraft, with Yuri Alexeyevich Gagarin onboard, was launched on April 12, 1961 at 9:07 a.m. Moscow time. The spacecraft, massing 4,725 kg, was put into an orbit with a perigee of 181 km, an apogee of 327 km, and an inclination of 65° by the three-stage launch vehicle (R-7 + block E) named Vostok. The Vostok spacecraft included a spherical descent vehicle (2.3 m in diameter and 2.46 tons in mass), a biconical instrumentation module (with a maximum diameter of 2.5 m and a mass of 2.265 tons), and the braking propulsion system developed by Isaev KB.

To return the descent vehicle with the cosmonaut to Earth the control system sent a command to the engine to provide a braking pulse; after that the spacecraft deorbited and then the descent vehicle separated from the instrumentation module and descended to Earth along the ballistic trajectory. At an altitude of 7 km the cosmonaut in a space suit left the descent vehicle using the ejection seat and then landed by parachute on his own. Having flown around the Earth in a matter of 108 minutes, Yu. A. Gagarin successfully descended to his native land.

On August 6, 1961 the Vostok-2 spacecraft, with cosmonaut G. S. Titov aboard, was launched. The cosmonaut was in space for an entire day.

The Vostok spacecraft program involved the launch of six manned spacecraft, including group flights of two pairs of spacecraft, and including the flight of the first woman-cosmonaut. The program was a success. On August 11 and 12, 1962 the Vostok-3 and Vostok-4 spacecraft were in space, and the Vostok-5 and Vostok-6 followed on June 16-19, 1963. The Vostok-6 spacecraft was piloted by Valentina Vladimirovna Tereshkova.

The experience accumulated in the development of the Vostok spacecraft was used to create the Voskhod three-man spacecraft (launched on October 12, 1964) and the Voskhod-2 two-man spacecraft. During the flight of Voskhod-2, on March 18, 1965, cosmonaut A. A. Leonov was the first in the world to egress into space. Upon completion of the program, the Vostok- and Voskhod-type spacecraft became technological history as new scientific and engineering ideas were pursued.

In 1957, work on the construction of automatic spacecraft designed for photography of Earth's surface was under way. In the course of this work, based on the Vostok spacecraft, the Zenit-2 unmanned spacecraft was designed, manufactured, tested and put into operation, and the Zenit-4 spacecraft design was developed. The first launch of the Zenit-2 spacecraft, on November 11, 1961, turned out to be a failure, caused by a rocket accident, but the second launch, on April 26, 1962, was a success.

Following a three-day flight, the spacecraft descent vehicle was returned to Earth. The Zenit-2 and Zenit-4 spacecraft were the beginning of a new trend in the creation of the national control aids using spacecraft. In 1964 the work on Zenit spacecraft was passed over to a subsidiary of KB which was headed by the OKB-1 former leading designer, D. I. Kozlov.

In 1961 the design work for creating the Molniya-1, the first communication satellite (active relay satellite), and construction of an experimental communications line based on it, was begun. Calculations showed that construction of a large number of comparatively simple and inexpensive ground receiving-transmitting stations and a relay satellite with a high-power radiated signal was more economically viable than constructing a central communication system and communicating with other stations via ground line networks. While developing the Molniya-1 satellite, the problem of satellite orientation was solved, and major advances were made in the designing of high power communication systems and their larger power supplies. On April 23, 1965, the first Molniya-1 satellite was launched into a highly elliptical orbit, and in 1968 a 24-hour communication system of three satellites was completed. Thereafter, work on communication satellites, as an independent development line in space technology, was passed over to the newly organized KB in Krasnoyarsk headed by S. P. Korolev's fellow campaigner M. F. Reshetnev.

Late in 1960, the Electron-1, Electron-2, Electron-3, and Electron-4 spacecraft were manufactured. These spacecraft included two satellites – E-I at 445 kg and E-II at 330 kg – which were injected into separate orbits by one launch-vehicle. The satellites were designed to explore the Van Allen radiation belt (regions of high-energy trapped plasma which come from the solar wind). The first pair of satellites was launched on January 30, 1964 and the second pair on July 11, 1964.

After launch of the first artificial Earth satellites, interplanetary stations to the Moon, Mars, and Venus, and flights of manned spacecraft in near-Earth orbit, the problem of constructing a new heavy launch vehicle was brought to the forefront. A launch vehicle capable of putting larger payloads into orbit was necessary in order to expand exploration of the planets and for creating a new generation of manned spacecraft capable of on-orbit docking. These are necessary for constructing a space system without which a wide study and exploration of space would be unthinkable.

In 1961, in parallel with the development of a new launch vehicle, the R-9 combat missile, with a launch mass of 81 tons and nose cone mass of 1.7-2.2 tons was manufactured at OKB-1 by the order of the Ministry of Defense. All prelaunch operations were fully automated. The flight range of the missile's nose cone was 12,500 km. Work on the creation of solid-propellant medium- and long-range missiles (RT-1 and RT-2) was also under way.

The N1 heavy launch vehicle was developed during the early 1960's. It was designed as a three-stage multipurpose rocket with a launch mass of 2,200 tons and a payload of 75 tons.

For a launch vehicle with this capability, special attention was paid to selection of the propellant components. A comprehensive comparison of characteristics of various pairs was conducted. As a result, a nontoxic, less expensive propellant pair – kerosene and liquid oxygen – was selected. It had the added benefit that both propellant components were already being produced. A large number of organizations were involved in development of the N1 rocket, fronted by the team led by N. D. Kusnetsov.

A series of rockets produced on the basis of the N1: the N11, using the second, third, and an additional fourth stage, had a launch mass of 700 tons and payload of 20 tons; the N111, using the third and an additional fourth stage, had a launch mass of 200 tons and payload of 5 tons. In conformance with the N1 project, a multi-engine system (24 engines in the first stage) was now used, the first in rocket building, so that the payload would be launched even if two pairs of engines failed. Because of the complexity of the multi-engine system, the rocket was equipped with the KORD special diagnostic system.

In May 1961 the USA proclaimed their Moon program and considered it their most important national task. Our country could not simply stand aside. In 1964, Korolev KB was entrusted by the government with the development of an analogous project. The "Moon race" had started. The mass of the N1 launch vehicle payload was increased initially up to 90 tons and then up to 95 tons. This increase was achieved through the installation of an additional six engines in the central part of the first stage and increasing of the propellant mass, raising the launch mass to 2,820 tons.

Concurrent with the work on the Moon program, development of the second generation of manned spacecraft, named Soyuz, was begun in 1962. On March 7, 1963, S. P. Korolev signed off on the design drawings for this spacecraft. In compliance with the requirements specified in 1965, a three-man spacecraft capable of performing a wide variety of tasks was designed, including: automatic and manual rendezvous and docking of spacecraft; performance of scientific and technological experiments; and testing of the autonomous navigation process.

The three-stage launch vehicle (R-7 + block I), subsequently called Soyuz, was used to put the Soyuz spacecraft into Earth orbit. The Soyuz spacecraft included the descent vehicle, crew habitation space, instrument assemblies, and strap-on modules.

The descent vehicle − about 3 tons with the thermal protection diameter of 2.2 m − was made in the shape of "a headlight" with an aerodynamic quality of 0.30 that, in combination with the descent control system microengines, provided a gliding descent with a g-load of no more than 4 g to a preselected landing area.

In January 1966, academician S. P. Korolev died. His successor, academician Vasiliy Pavlovich Mishin, continued the work on the development of the N1 rocket and the Soyuz spacecraft.

On November 28, 1966, the Soyuz spacecraft flight testing began in unmanned mode. Following the second unmanned flight (February 7, 1967), on April 23, 1967, the Soyuz-1 with pilot-cosmonaut V. M. Komarov aboard was launched. The flight ended in tragedy. Because of a landing system failure the cosmonaut perished. Following improvements, testing of the unmanned spacecraft was repeated. The Cosmos-186 and Cosmos-188 unmanned spacecraft, which on October 30, 1967 were the first in the world to dock in orbit in an automatic mode, were launched. The Cosmos-212 and Cosmos-213 unmanned spacecraft repeated automatic docking in orbit.

The five unmanned spacecraft flights (including Cosmos-238) confirmed the validity of the adopted solutions. The decision to perform a manned flight was again made. Cosmonaut G. T. Beregovoy flew in space aboard the Soyuz-3 spacecraft. His spacecraft was launched on October 26, 1968, following the Soyuz-2 unmanned spacecraft. During this flight the spacecraft automatic rendezvous and manual berthing were tested.

On January 15, 1969, the Soyuz-4 (cosmonaut V. A. Shatalov) and Soyuz-5 (cosmonauts B. V. Volynov, A. S. Eliseev, E. V. Khrunov) manned spacecraft docked in orbit, constructing an experimental space station of 12.924 tons. Two cosmonauts in space suits passed from one spacecraft to the other through space. In June 1970, cosmonauts A. G. Nikolaev and V. I. Sevastianov performed a long-duration flight (17.7 days) on board the Soyuz-9 spacecraft. A Soyuz spacecraft transport modification, and later its modification with the androgynous periphery docking unit, was then put under development.

Beginning in 1965, an additional modification of the Soyuz spacecraft designed for the Moon fly-around was under development. It was planned for the Soyuz spacecraft to be launched by the Proton four-stage launch vehicle. Booster D, developed by TsKBEM (Korolev KB was so named), was used as the fourth stage of the Proton launch vehicle – the first upper stage providing multiple engine ignitions in space. It was equipped with a TsKBEM-designed closed-cycle engine with 8.5 tons thrust. Using liquid oxygen and kerosene, the engine had a high specific impulse (349 kg.f-s/kg). On March 10, 1967, the unmanned launches of the 7K-L1 spacecraft of this series, named Zond, began. During the period 1968-1970 these unmanned spacecraft, from Zond-5 to Zond-8, flew around the Moon.

After Moon fly-around and photography, the first of these spacecraft, Zond-5, splashed down in the Indian ocean. For a number of reasons the Moon fly-around by a two-man crew on board the 7K-L1 manned spacecraft did not take place.

During subsequent years, the D-booster was improved and called DM. In 1974-1993, the DM-booster, coupled with the Proton launch vehicle, provided launching of over 130 space objects of the Cosmos, Venera, Raduga, Ekran, Gorizont, Vega, Fobos series, etc.

In late 1969, on a basis of the scientific and technological products available at TsKBEM and subsidiary TsKBM's (hereafter KB Salyut), the immediate development of an orbital station was begun. The orbital and core module body created for the Almaz manned station formed the station basis. Structurally, the station consisted of a work module with zones of large (4.15 m) and small (2.9 m) diameters, and transfer and service modules. The volume of the first station habitation module was 90 m^3, and the mass of the scientific equipment was 1.2 tons.

On April 19, 1971, the world's first orbital station, named Salyut, was put into Earth orbit by the Proton three-stage launch vehicle. The Soyuz-10 spacecraft was to deliver the crew to the station, but because of a failure in the mechanical docking system, the crew could not transfer to the station. On June 8, 1971, the first crew, including G. T. Dobrovolsky, V. N. Volkov and

V. I. Patsaev, arrived at the station on board the Soyuz-11 spacecraft and worked there for 22 days, performing a large number of investigations. However, during the descent phase while returning to Earth, a premature opening of the ventilation system pyrotechnic valve occurred resulting in the tragic deaths of the cosmonauts.

After this the station made a flight in automatic mode. Scientific and technical investigations, and control of the systems, structure and scientific equipment under long-duration flight conditions were performed. The Salyut station stayed in near-Earth orbit for about 6 months (until November 11, 1971).

On May 11, 1973, the next orbital station – Cosmos-557 – was put into orbit. Because of the abnormal operation of the ionic orientation system, the flow rate of the working medium in the actuators system considerably exceeded design values. Station orbit correction was impossible and within 12 days the station ceased to operate.

The next orbital station – Salyut-4, developed by TsKBEM and KB Salyut – was launched on December 26, 1974 and was in orbit until February 3, 1977. Two expeditions, of 28 and 63 days duration, worked aboard the station. The crews on board conducted integrated scientific and technological experiments. The checkouts of the station's structure, units and systems under conditions of a long-duration flight (resource tests) were of considerable importance.

In 1973 TsKBEM and KB Salyut began a joint development of a new generation station. Its most distinctive feature was a second docking unit. While developing the station special attention was paid to its maintainability in order to increase its lifetime.

Late in 1968, the assembly of the first N1 launch vehicle was completed, and on February 21, 1969 the first launch took place. Its flight duration was only 68.7 seconds because of a fire in the aft section of the first stage, causing the KORD system to cut off all engines. For that first launch, the N1 launch vehicle mass was 2,735 tons, with a first stage thrust of 4,500 tons, and payload of about 70 tons.

During the second N1 launch, on July 3, 1969, the launch vehicle had an accident during the first seconds of flight and the rocket fell down onto the launching pad. Subsequently, the N1-L3 flight tests were protracted, time being necessary to clarify the causes of the failures and adopt measures for their elimination.

On July 24, 1969, the crew of the U.S. Apollo-11 spacecraft returned to Earth after landing on the Moon's surface and political interest in our Moon program vanished.

The development of the booster and spacecraft for the Moon program had been completed. The operational capability of the Lunar spacecraft was checked out in near-Earth orbit as a part of the T2K unmanned experimental spacecraft which was launched by the Soyuz launch vehicle on November 24, 1970 (Cosmos-379), February 26, 1971 (Cosmos-398) and August 12, 1971 (Cosmos-434).

The third (June 27, 1971) and fourth (November 23, 1972) launches of N1-L3 were not successful. In December 1972, the USA completed their Moon program with the Apollo-17 flight, which determined the fate of the N1 rocket.

In May of 1974, NPO Energia, the main part of which became TsKBEM, was headed by academician Valentin Petrovich Glushko. By that time the preparation of the Soyuz-Apollo flight had been completed. The program director of the Soviet part was K. D. Bushuev. Two Soyuz spacecraft and four crews were in preparation for the flight. In July of 1975, the Soyuz-19 and U.S. Apollo spacecraft docked in orbit. Soviet cosmonauts A. Leonov and V. Kubasov shook hands and exchanged pennants with the U.S. astronauts T. Stafford, V. Brand, and D. Slayton and they performed joint experiments. The flight was successfully completed with the cosmonaut's landing.

The extra spacecraft that wasn't used by the Soyuz-Apollo program was reoriented for use in the Intercosmos program whose purpose was to test and improve scientific and technological methods for studying Earth's geological and geophysical characteristics from space in the interests of the national economy and environmental monitoring. For this purpose the special photocompartments with a multi-zonal photographic apparatus (MKF-6) developed by the USSR and GDR was installed on board. The Soyuz-22 spacecraft flight was conducted in September 1976.

In February 1976, NPO Energia was charged with the development of a reusable rocket and space system including the Energia launch vehicle and Buran orbital vehicle. This system was created to counterbalance the U.S. Space Shuttle transportation system so as to maintain parity with the US militarily and with respect to subsequent space exploration. An important difference between this and earlier programs was that the heavy-lift launch vehicle and the orbital spacecraft were being created separately.

The Energia launch vehicle, with a launch mass of 2,400 tons and initial thrust of 3,550 tons, is a two-stage rocket integrated in a single package. The first stage consists of four side boosters with a four-chamber liquid-fuel engine burning liquid oxygen and hydrocarbon in each booster. The second stage is the vehicle's central module with four liquid-fuel rocket engines burning liquid oxygen and liquid hydrogen.

After completion of thorough ground testing, the first launch of the Energia rocket, with the "Skif-DM" (or "Polus") spacecraft designed at KB Salyut, was performed on May 15, 1987.

The Buran orbiter was developed in parallel with the launch vehicle. The orbiter was being tested under flight conditions with the use of a prototype spacecraft. Additional engines were installed on the prototype orbiter. On November 10, 1985, it performed its first flight over Zhukovsky town. Development of the orbiter systems and on-board automatic equipment, including software, had also been proceeding. The first flight of the orbiter was planned to be unmanned. At last, on November 15, 1988 at 6:00 a.m. Moscow time, the Energia-Buran system made its first flight.

A combined propulsion system of NPO Energia design was installed in the Buran orbiter. It included engines for orbital maneuvering, control and precise orientation. Oxygen and synthetic hydrocarbon fuel, which all engines burned, were contained in common propellant tanks.

After completing a two-circle orbital flight, the Buran orbiter performed an automatic landing on an airfield not far from the launch site. The automatic landing system provided landing accuracy within centimeters of the design prediction. The flight duration was 205 minutes.

On September 29, 1977, a new stage in manned cosmonautics was opened with the Salyut-6 station launch. Salyut-6 was a new generation station equipped with two docking units. The station was first visited by the crew of the Soyuz-26 spacecraft launched on December 11, 1977. Delivery of propellants for the propulsion system and different cargoes to the station was provided by Progress unmanned cargo spacecraft (the first launch was made on January 20, 1978) created on the Soyuz spacecraft basis.

The first international crew, consisting of spacecraft commander A. A. Gubarev and cosmonaut-researcher V. Remek (ChSSR), was delivered to the station on March 3, 1978 by the Soyuz-28 spacecraft (launched on March 2, 1978). They performed scientific and technical research during their stay on board the station.

On December 16, 1979, a new Soyuz T unmanned transport spacecraft, developed on the basis of the Soyuz spacecraft, was launched. New onboard systems, including systems for radio communication, attitude control, motion control and an onboard computer complex, were installed aboard the Soyuz T spacecraft. On December 19, 1979, the spacecraft was docked to the Salyut-6 station and remained docked, being tested as a part of the station complex, for more than 100 days. A manned version of the Soyuz T spacecraft became the main transport vehicle for delivering cosmonauts to the orbital stations. Soyuz spacecraft T-2 delivered a crew to the station on June 6, 1980.

Between 1977 and 1981, 16 crews carried out work aboard the Salyut-6 station (it deorbited on July 29, 1982), and the total stay duration was 676 days. During that time unique research was performed in astrophysics, geophysics, substance structure, and on the effects of long-term flight conditions on the human organism. Additionally, a survey of Earth's natural resources; ecological monitoring of the Earth's surface, lakes, rivers, and atmosphere; production of new materials and highly effective biological substances; and EVA's were performed.

On April 19, 1982, the Salyut-7 station was put into orbit. The crew was delivered to the station by the Soyuz T-5 spacecraft launched on May 13, 1982. Ten crews worked aboard the Salyut-7 station, continuing research work begun by cosmonauts on board the Salyut-6 station. The total flight duration in the manned mode was about 800 days. Eleven cargo spacecraft of the Progress-series and two logistics spacecraft of 20-ton class – Cosmos-1443 and Cosmos-1686 (jointly designed by KB Salyut and TsKBM) – delivered propellants and cargoes to the station. In October of 1984, the Salyut-7 station, with the docked transport logistics spacecraft Cosmos-1686, was transferred into a 480 km orbit to perform prolonged life tests of the complex equipment and systems in automatic mode.

Early in 1985, the power supply system of the Salyut-7 station failed. The station's orientation was disturbed and it no longer responded to Control Centre commands. In June of 1985, to restore the station's serviceability, the Soyuz T-13 spacecraft was launched, which docked successfully to the station in the manual control mode. The cosmonauts restored the station's operability. The Salyut-7 / Cosmos-1686 complex terminated its functioning on February 7, 1991.

The accumulated experience of the Salyut-6 and Salyut-7 stations made it possible to proceed to creation in orbit of a permanent manned complex with specialized orbital modules for scientific and national economic purposes. The Mir orbital station – yet another new generation station – formed the core module of a permanent complex. The station was equipped with a new docking system and six docking units. The core module and add-on modules of the complex were developed jointly with KB Salyut.

On February 20, 1986, the Mir core was put into orbit. On March 15, 1986, the Soyuz T-15 spacecraft delivered the first crew to the station. The crew stayed aboard the station until May 5, then the Soyuz T-15 spacecraft, with the crew on board, was undocked and performed the world's first orbital transfer to the Salyut-7 station. The crew operated on board the Salyut-7 station for over 25 days, and then, on June 26, 1986, the Soyuz T-15 spacecraft returned them to the Mir station, bringing along about 400 kg of scientific equipment from Salyut-7 for further use on the Mir complex.

To deliver crews to the multipurpose manned complexes of the modular type, a modified spacecraft – Soyuz TM – was developed. The Soyuz TM included new systems, among them, systems for rendezvous, radio communication, emergency rescue, and a new combined propulsion system. On May 21, 1986, an unmanned Soyuz TM spacecraft docked to the Mir station for complex experimental tests in automatic flight with the station.

On February 6, 1987, the Soyuz TM-2 spacecraft delivered a new crew to the Mir station and on March, 31, the first scientific (astrophysics) module – Kvant – was docked to the station.

Since 1989, NPO Energia has been headed by Juri P. Semenov, and the manned programs are being further developed. Continuing on, the Kvant-2 add-on module (December 6, 1989) and Kristall technological module (May 31, 1990) were docked to the station, and the Mir station program became goal-oriented.

Developments in orthopedic prosthetics and the creation of different consumer products were added to the main activities of NPO Energia. Within a short period of time the NPO Energia specialists, engaged in space subjects, mastered the production of prostheses, which are highly competitive with the best foreign offerings.

The search for new, even more effective launch vehicles, and the planning of more manned programs proceeded vigorously in the field of space exploration. A ballistic recovery capsule was developed for installation in the Progress M transport cargo spacecraft. At the completion of a mission, during descent, this capsule separates from the spacecraft and delivers the research results to the ground. The first ballistic capsule was delivered to the Mir complex by the Progress M-5 cargo spacecraft on September 27, 1990, and was returned to the region of the descent vehicle's landing site on November 28, 1990.

International co-operation has been continually pursued. The Mir space station remained in orbit for more than 15 years until it deorbited in March of 2001. The Mir station clearly confirmed the efficiency and practical return of the module-type space station.

Despite economic difficulties, the NPO Energia staff retains its creative potential and does its best to continue the development of national rocket-space technology, being true to Korolev's precept – "so little is achieved, so much is to be done."

Based on the Energia launch vehicle, NPO Energia has created a configurable series of launch vehicles. By selecting the set of side boosters to be used and then effecting standard modifications to the central module, the series of launch vehicle permutations achievable make it possible to effectively put into orbit payloads of widely differing masses – a light class launch vehicle is capable of putting into near-Earth orbit a payload of up to 5 tons, while a superheavy-class launch vehicle can lift to orbit up to 200 tons of payload. The Energia-M launch vehicle, capable of lifting up to 34 tons, is of particular interest in this series. Because NPO Energia offers this configurable series of launch vehicles, cosmonauts get a unique system solution for each mission, tailored specifically to the mission, for payloads from light to superheavy.

Availability of these practical and efficient launch vehicles provide Russia with the ability to solve all its national economic and scientific problems, to offer launch vehicles to the international marketplace, and to extend international cooperation when performing joint space programs.

Only through the use of the Energia launch vehicle can we address most efficiently the problems of mankind that can be solved only by the exploration and exploitation of space. The Energia launch vehicle provides effective and global solutions for tasks pertaining to communication, broadcasting, and ecology that require the use of large space platforms, and exploration of the Moon, Mars, and the Solar System.

The Mir permanent orbital space station has played a specific and vital role in the furtherance of space technology. The experience gained on Mir will help us to define an optimum program of space exploration. Only during long-term manned flights can fundamental research be conducted in astrophysics, geophysics, ecological monitoring of the Earth's surface, lakes, rivers and atmosphere, and the Earth's natural resources. As well, production can be developed for valuable materials and biological commodities whose unique properties are only available from manufacturing in space.

Since July 1994, NPO Energia has been called S P. Korolev Space Corporation Energia (RSCE).

RSCE has maintained that creation of orbital space stations should become an international affair and has considered a number of proposals on co-operation. The well-developed Soyuz TM spacecraft is ideally suited as an ACRV for any international programs, including international space stations. NPO Energia (RSCE) is currently a prime contractor and Russia's main contributor for the International Space Station project.

The offer of RSCE participation is open to everybody, and the results of this activity may be used by any organization in any country.

RSCE stands ready to provide launch vehicles, spacecraft and orbital stations for investigations and explorations in mutually beneficial space programs.

Sergey Pavlovich Korolev
The founder of practical cosmonautics.
Chief Designer of the first rocket / space systems.
The founder and first manager of OKB-I (1946-1966)

The Council of Chief Designers – consisting of M. S. Rjazansky, N. A. Piljugin, S. P. Korolev, V. P. Glushko, V. P. Barmin, and V. I. Kuznetsov – was organized on Korolev's initiative. Complex problems in the development of specific areas of rocket / space technology were discussed by the Council.

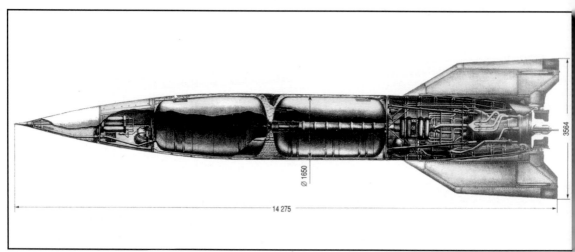

The first Russian rocket, the R-1, was designed under the leadership of S. P. Korolev. The R-1 rocket complex, put into operation in 1950, included both engineering and launch facilities. The R-1 rocket was manufactured in a series of variations, each specific to a particular type of task.

The engineering facilities for the R-1 rocket.

Electrical supply unit

Accumulator/Battery vehicle

Autonomous test vehicle

Horizontal test vehicle

Rocket on ground trolley

Pneumatic hoist

Tent (hangar)

Mobile Electrical power stations

Power supply unit

Compressor station

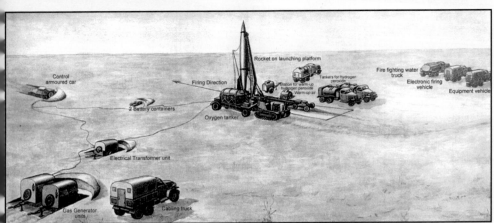

The launch facilities for the R-1 rocket.

Control armoured car

Firing Direction

Rocket on launching platform

Station for warm-up hydrogen peroxide

Tankers for hydrogen peroxide

Warm-up air

Fire fighting water truck

Electronic firing vehicle

Equipment vehicle

2 Battery containers

Oxygen tanker

Electrical Transformer unit

Gas Generator units

Cabling truck

Completion of the R-1 rocket launch preparation.

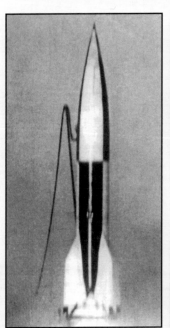

The launch of the R-1 rocket.

The R-1 rocket in flight

Rockets to investigate the upper atmosphere on the R-1 rocket base

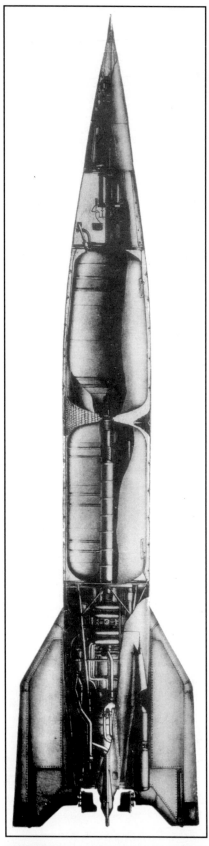

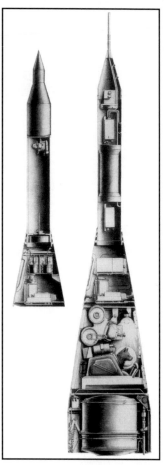

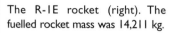

The R-1E rocket payload module. The recoverable payload module mass was 760 kg.

The R-1A rocket (left). The first rocket that delivered scientific equipment in recoverable containers (seen in the area of the stabilizers) into the upper atmosphere.

The R-1E rocket (right). The fuelled rocket mass was 14,211 kg.

Installation of an instrument container into a payload carrying mortar.

Recoverable instrument container after flight.

Landing of the rocket's payload upon flight completion.

The R-1D rocket payload module.

The R-1D rocket (left).

The R-1D rocket on the launch pad with the carriage lowered.

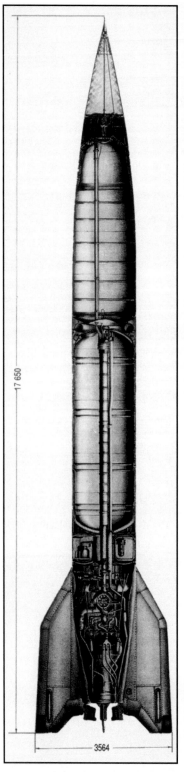

17 650

3564

The engineering facilities for the R-2 rocket.

The launch facilities for the R-2 rocket.

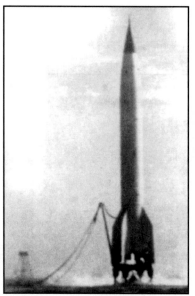

The R-2 rocket engine firing.

The R-2 rocket in flight.

The R-2 rocket. This rocket had a separable payload module.
Regular launches of the rocket began on October 26, 1950.
The R-2 was developed in the shortest possible time owing
to the use of parts and rigging from the R-1 rocket design.

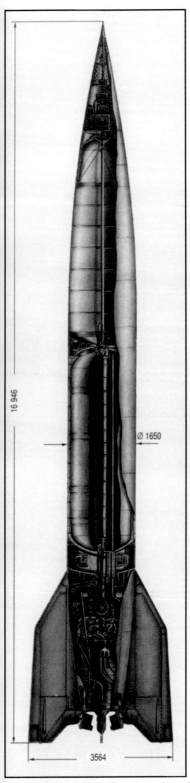

The R-2A rocket – designed on the basis of the R-2, to investigate the upper atmosphere – before launching

The R-2A rocket in flight.

The R-2A rocket payload module (right).

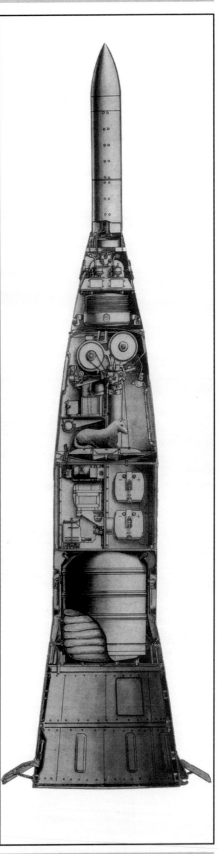

The R-2E rocket. The first launch of the R-2E experimental rocket was performed on September 21, 1949. Rocket launches were performed to test the serviceability of the R-2E rocket's systems.

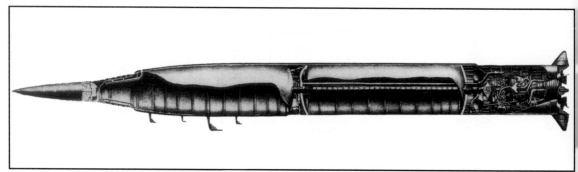

The first strategic rocket, the R-5. The fuelled rocket mass was 28,570 kg.

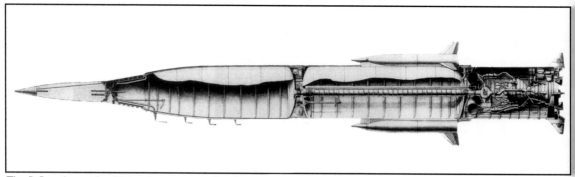

The R-5 rocket with additional strap-on warheads.

The engineering facilities for the R-5 rocket.

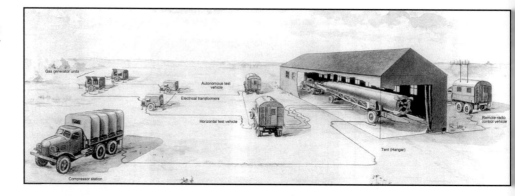

The launch facilities for the R-5 rocket.

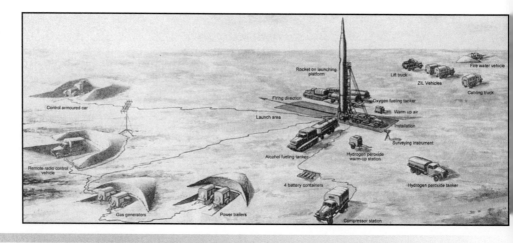

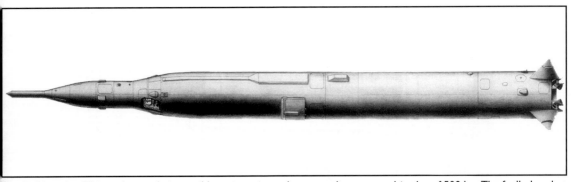

The R-5A rocket. This rocket made it possible to investigate the atmosphere up to altitudes of 500 km. The fuelled rocket mass was 29,314 kg.

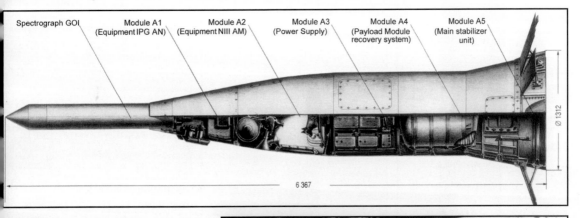

Spectrograph GOI | Module A1 (Equipment IPG AN) | Module A2 (Equipment NIII AM) | Module A3 (Power Supply) | Module A4 (Payload Module recovery system) | Module A5 (Main stabilizer unit)

Ø 1312

6 367

The R-5A rocket payload module. The recoverable payload module mass is 1,350 kg.

The rocket payload module upon separation during integrated tests

The R-5A rocket before launch.

The recovered payload module after landing.

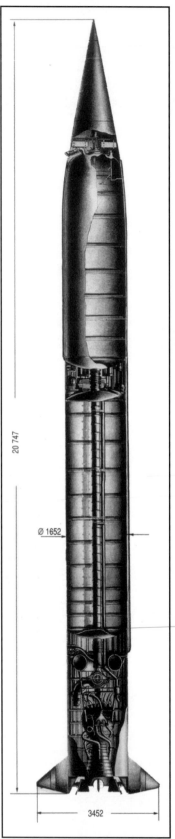

Maximum range under normal firing conditions (t=15°C and p=760 mm barometric) without calculating for Earth's rotation, KM	1200
Velocity at Engine cut-off M/S	3016
Peak Trajectory	304
Flight time to target, sec	637
Rocket lift-off mass, KG	28 610
Dry rocket mass, KG	4390
Fuel mass, Hydrogen peroxide and air, KG	24 500
including:	
Oxygen, KG	13 990
Alcohol, KG	10 010
Engine thrust at sea level, KGF	43 860
Specific impulse at sea level, KGF.S/KG	219,3
Engine burn time, sec	115,4

Installation of the R-5M rocket onto the launch pad.

The R-5M rocket engine firing.

R-5M rocket launch processing (above).

The R-5M strategic rocket with a nuclear charge (left).

Transportation of the R-5V rocket and its installation on the launch pad. R-5V launches were performed until 1975 as part of the vertical program.

A *Pravda* newspaper report on atmospheric investigation using rockets.

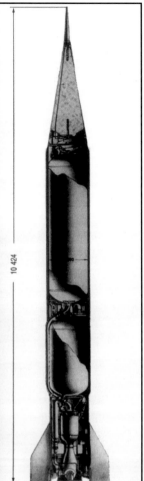

Transportation of the R-11 rocket.

Maximum straight line range, KM	270
Mass, KG:	
fueled rocket	5350
dry rocket	1645
main unit	700
Mass of fuel components and compressed air, KG including:	3700
Oxidizer AK-20, KG	2900
Fuel T-1, KG	705
Thrust at sea level, KGF	8300
Specific impulse at sea level, KGF-S/KG	219

Installation of the R-11 rocket on the launch pad.

The R-11 was the first operative tactical rocket to burn a storable propellant. The R-11 was highly mobile. The rocket's launch mass was 5,350 kg.

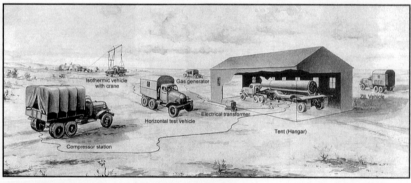

The engineering facilities for the R-11 rocket.

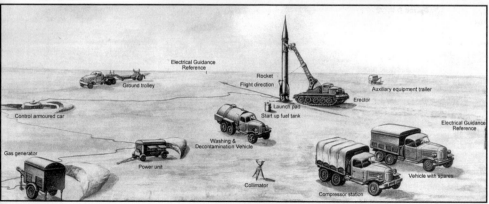

The launch facilities for the R-11 rocket.

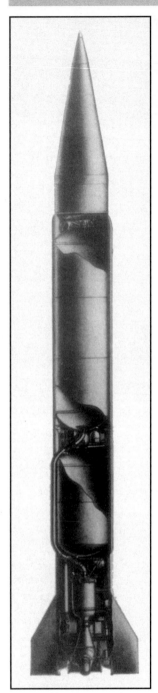

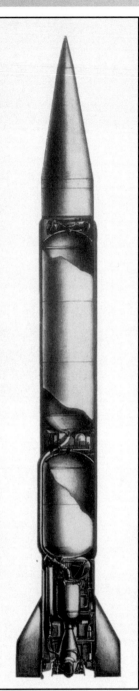

The R-11FM rocket on a submarine, ready for launch.

The R-11FM rocket launch from a submarine.

The R-11M rocket with a nuclear charge.

The sea-based R-11FM rocket. This rocket was launched from a submarine in the above-water position.

The R-11FM rocket immediately following launch from the submarine.

Design and development of launch vehicles

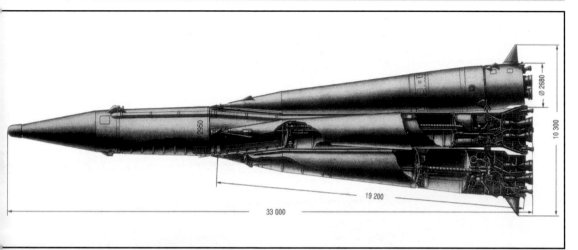

The R-7 intercontinental rocket. This was the world's first rocket capable of delivery of a nuclear warhead to any point in potential enemy territory. The Earth's first artificial satellites were launched using this rocket.

Rocket assembly and systems checks were performed in the stationary assembly-test building. The four-chamber main engines and control engines (a four-chamber engine in the core, and a two-chamber on the side module) can be seen.

The R-7 rocket in flight. The first successful launch was performed on August 21, 1957.

The R-7 rocket before launch (May 15, 1957).

The R-7 rocket was launched from the stationary launch facility – a complex engineering facility.

Onset of Space Era

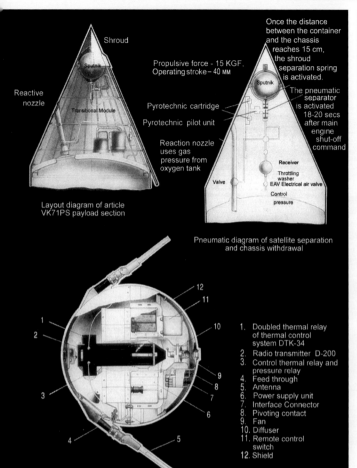

Shroud

Sputnik

Reactive nozzle

Transitional Module

Layout diagram of article VK71PS payload section

Propulsive force - 15 KGF. Operating stroke – 40 мм

Pyrotechnic cartridge

Pyrotechnic pilot unit

Reaction nozzle uses gas pressure from oxygen tank

Once the distance between the container and the chassis reaches 15 cm, the shroud separation spring is activated.

The pneumatic separator is activated 18-20 secs after main engine shut-off command

Sputnik

Receiver

Throttling washer
EAV Electrical air valve

Valve

Control pressure

Pneumatic diagram of satellite separation and chassis withdrawal

12
11
10
9
8
7
6
5
4
3
2
1

1. Doubled thermal relay of thermal control system DTK-34
2. Radio transmitter D-200
3. Control thermal relay and pressure relay
4. Feed through
5. Antenna
6. Power supply unit
7. Interface Connector
8. Pivoting contact
9. Fan
10. Diffuser
11. Remote control switch
12. Shield

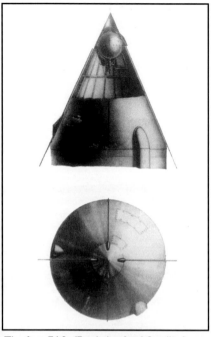

The first EAS (Earth Artificial Satellite) was mounted under the launch vehicle's payload shroud.

On October 3, 1957, the world learned the Russian word "Sputnik." On that day TASS informed the world of the launch of the first artificial satellite. Sputnik massed 83.6 kg and was the first man-made object to orbit the Earth.

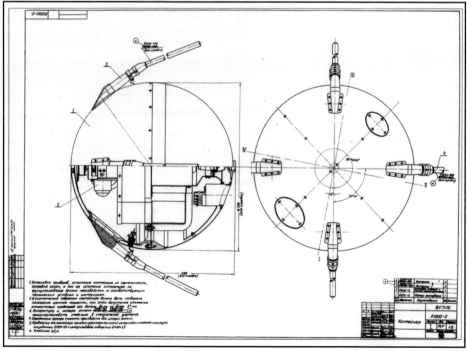

The container for the first EAS.

Lift-off mass, KG	267
Engine thrust at launch, TF	398
Specific impulse KGF-S/KG	250
Payload Mass, T	1327
Altitude, M	29,167

The Sputnik launch vehicle was designed on the basis of the R-7 rocket.

The state commission on Earth's first artificial satellite launch (first row, from left to right): G. R. Udarov, I. T. Bulychev, A. G. Mrykin, M. V. Keldysh, S. P. Korolev (technical manager), V. M. Rjabikov (chairman of the commission), M. I. Nedelin, G. N. Pashkov, V. P. Glushko, V. P. Barmin, (second row) M. S. Rjazansky, K. N. Rudnev, N. A. Piljugin, S. M. Vladimirsky, and V. I. Kuznetsov.

The launch vehicle with the first EAS immediately following lift-off.

Сообщение ТАСС

В течение ряда лет в Советском Союзе ведутся научно-исследовательские и опытно-конструкторские работы по созданию искусственных спутников Земли.

Как уже сообщалось в печати, первые пуски спутников в СССР были намечены к осуществлению в соответствии с программой научных исследований Международного геофизического года.

В результате большой напряженной работы научно-исследовательских институтов и конструкторских бюро создан первый в мире искусственный спутник Земли. 4 октября 1957 года в СССР произведен успешный запуск первого спутника. По предварительным данным, ракета-носитель сообщила спутнику необходимую орбитальную скорость около 8 000 метров в секунду. В настоящее время спутник описывает эллиптические траектории вокруг Земли и его полет можно наблюдать в лучах восходящего и заходящего Солнца при помощи простейших оптических инструментов (биноклей, подзорных труб и т. п.).

Согласно расчетам, которые сейчас уточняются прямыми наблюдениями, спутник будет двигаться на высотах до 900 километров над поверхностью Земли; время одного полного оборота спутника будет 1 час 35 минут, угол наклона орбиты к плоскости экватора равен 65°. Над районом города Москвы 5 октября 1957 года спутник пройдет дважды — в 1 час 46 мин. ночи и в 6 час. 42 мин. утра по московскому времени. Сообщения о последующем движении первого искусственного спутника, запущенного в СССР 4 октября, будут передаваться регулярно широковещательными радиостанциями.

Спутник имеет форму шара диаметром 58 см и весом 83,6 кг. На нем установлены два радиопередатчика, непрерывно излучающие радиосигналы с частотой 20,005 и 40,002 мегагерц (длина волны около 15 и 7,5 метра соответственно). Мощности передатчиков обеспечивают уверенный прием радиосигналов широким кругом радиолюбителей. Сигналы имеют вид телеграфных посылок длительностью около 0,3 сек., с паузой такой же длительности. Посылка сигнала одной частоты производится во время паузы сигнала другой частоты.

Научные станции, расположенные в различных точках Советского Союза, ведут наблюдение за спутником и определяют элементы его траектории. Так как плотность разреженных верхних слоев атмосферы достоверно неизвестна, в настоящее время нет данных для точного определения времени существования спутника и места его вхождения в плотные слои атмосферы. Расчеты показали, что вследствие огромной скорости спутника в конце своего существования он сгорит при достижении плотных слоев атмосферы на высоте нескольких десятков километров.

В России еще в конце 19 века трудами выдающегося ученого К. Э. Циолковского была впервые научно обоснована возможность осуществления космических полетов при помощи ракет.

Успешным запуском первого созданного человеком спутника Земли вносится крупнейший вклад в сокровищницу мировой науки и культуры. Научный эксперимент, осуществляемый на такой большой высоте, имеет громадное значение для познания свойств космического пространства и изучения Земли как планеты нашей солнечной системы.

В течение Международного геофизического года Советский Союз предполагает осуществить пуски еще нескольких искусственных спутников Земли. Эти последующие спутники будут иметь увеличенные габарит и вес и на них будет проведена широкая программа научных исследований.

Искусственные спутники Земли проложат дорогу к межпланетным путешествиям и, по-видимому, нашим современникам суждено быть свидетелями того, как освобожденный и сознательный труд людей нового, социалистического общества делает реальностью самые дерзновенные мечты человечества.

The prototype of the first EAS and its shroud in the RSCE museum.

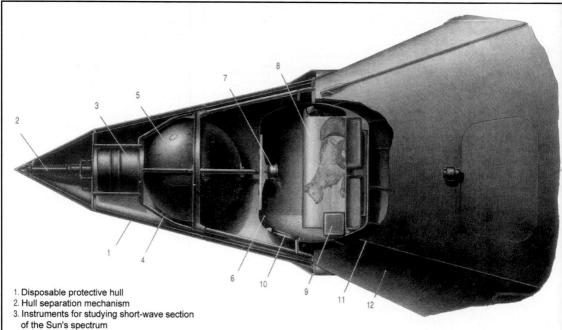

1. Disposable protective hull
2. Hull separation mechanism
3. Instruments for studying short-wave section
   of the Sun's spectrum
4. Instrument frame
5. Spherical container with radio transmitter
6. Pressurized cabin with experimental animal
7. Fan
8. Air scrubbing unit
9. Food trough
10. Window
11. Antenna
12. Transfer Module

The second EAS, which massed 508.3 kg. The dog Laika
was the passenger aboard the satellite.

Laika before boarding the special EAS compartment.

The prototype of the second EAS in the RSCE museum.

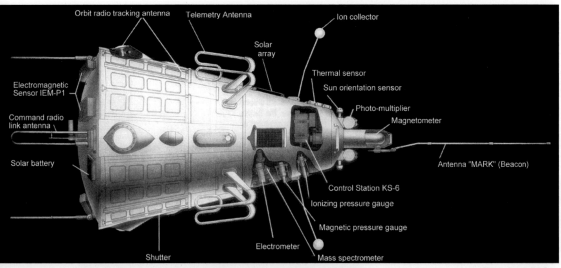

Orbit radio tracking antenna — Telemetry Antenna — Ion collector — Solar array — Thermal sensor — Sun orientation sensor — Photo-multiplier — Magnetometer — Electromagnetic Sensor IEM-P1 — Command radio link antenna — Solar battery — Antenna "MARK" (Beacon) — Control Station KS-6 — Ionizing pressure gauge — Magnetic pressure gauge — Shutter — Electrometer — Mass spectrometer

The third EAS.

Mating of the third EAS to its launch vehicle.

The third EAS frame with instruments and power supply units.

The third EAS body in the RSCE museum.

Onset of Flights to the Moon

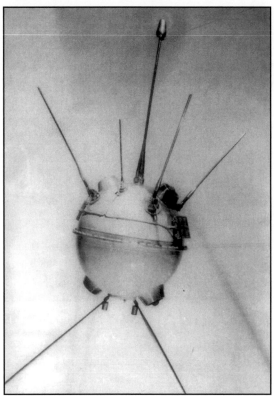

Mankind's dream had come true. The Earth's first messenger to the moon – the Mechta interplanetary station (Luna-1) flew at a distance of 5-6 thousand km from the Moon and then became a satellite of the Sun.

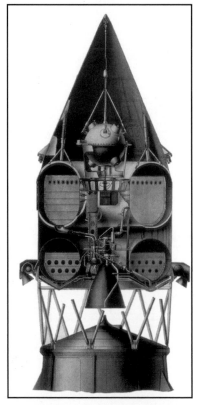

Accommodation of the lunar interplanetary station under the E rocket module payload shroud.

The R-7 rocket with the E module and Luna-1 interplanetary station.

The Luna-2 interplanetary station (above) and the prototype of the Luna-2 in the RSCE museum (right).

Pennants delivered to the Moon by the Luna-2 interplanetary station.

The prototype of the Luna-3 interplanetary station in the RSCE museum.

The Luna-3 interplanetary station.

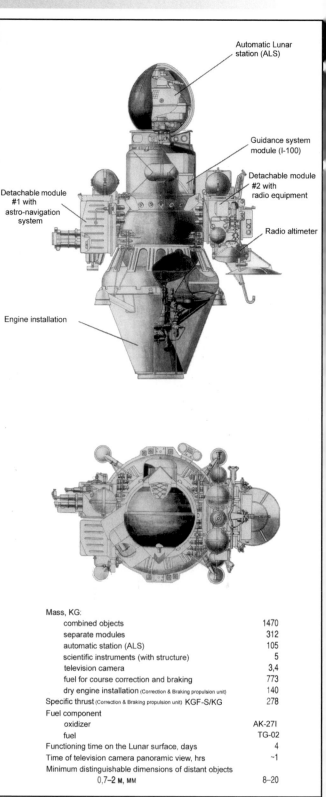

Mass, KG:	
combined objects	1470
separate modules	312
automatic station (ALS)	105
scientific instruments (with structure)	5
television camera	3,4
fuel for course correction and braking	773
dry engine installation (Correction & Braking propulsion unit)	140
Specific thrust (Correction & Braking propulsion unit) KGF-S/KG	278
Fuel component	
oxidizer	AK-27I
fuel	TG-02
Functioning time on the Lunar surface, days	4
Time of television camera panoramic view, hrs	~1
Minimum distinguishable dimensions of distant objects	
0,7–2 м, мм	8–20

The general view of the Luna-9 automatic interplanetary station.

The prototype of the automatic lunar station in the RSCE museum.

The world's first closed-loop liquid rocket engine had a thrust of about 7 tons and was developed at Korolev's KB. The engine was installed on the L booster of the Molniya four-stage launch vehicle.

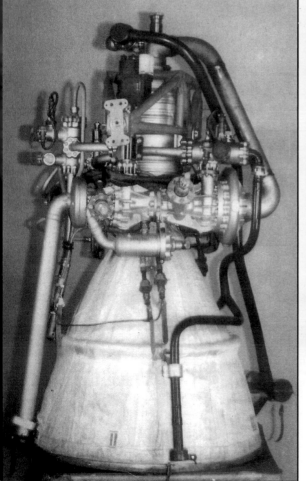

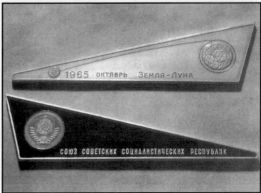

Pennants delivered by automatic interplanetary stations to the Moon.

Transportation of a four-stage launch vehicle (R-7 plus rocket unit I plus booster L) with an interplanetary station.

Installation of the four-stage launch vehicle with an automatic interplanetary station onto the launch pad.

First vehicles to investigate Venus and Mars

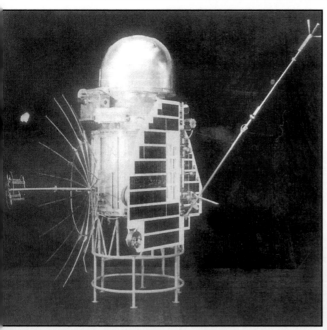

Venera-1 (1VA).

Venera-2 (3MV-4 No. 4).

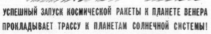

A Pravda newspaper report on the launch of the Venera-1 automatic interplanetary station.

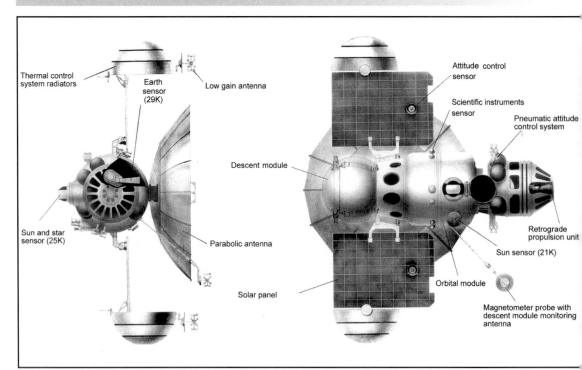

Zond-1 (3MV-1 No.4).

Venera-2 (3MV-4 No. 4).

Mars-1 (2MV-4 No. 4).

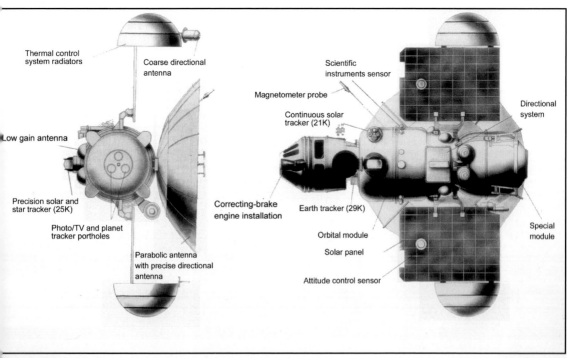

Zond-2 (3MV-4 No. 2).

Labels in the figure:
- Thermal control system radiators
- Coarse directional antenna
- Scientific instruments sensor
- Magnetometer probe
- Continuous solar tracker (21K)
- Directional system
- Low gain antenna
- Precision solar and star tracker (25K)
- Photo/TV and planet tracker portholes
- Correcting-brake engine installation
- Earth tracker (29K)
- Special module
- Parabolic antenna with precise directional antenna
- Orbital module
- Solar panel
- Attitude control sensor

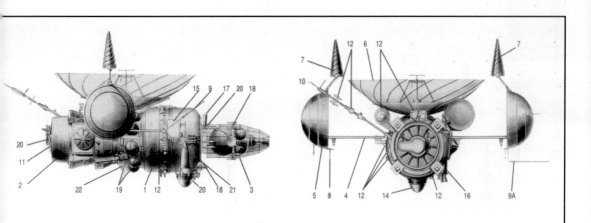

1. Pressurized orbital module
2. Special pressurized module (photo-module)
3. Correcting brake engine installation
4. Solar panel
5. Thermal control system radiators
6. High gain parabolic antenna
7. Low gain antenna
8. Low gain antenna
9. Meter wave-band transmitter antenna
9A. Meter wave-band receiver antenna
10. Omni-directional emergency radio antenna

11. Photo/TV and planet tracker portholes
12. Science instruments sensor
14. Precision solar and star tracker
15. Contingency radio link
16. Continuous solar tracker
17. Parabolic antenna Earth tracking sensor
18. Attitude control system nozzles
19. Attitude control system compressed gas tanks
20. Attitude sensor shutters
21. Non-precision sun tracker
22. Sun tracker

Total mass of object, KG	910
Mass radio instrumentation, KG	160
Mass Correcting Brake Engine, KG	68

Mars-1 (2MV-4 No. 4).

Venera-3 (3MV-3 No. 1).

Pennants delivered by the Venera- automatic station to the surface c Venus.

Onset of Manned Flight

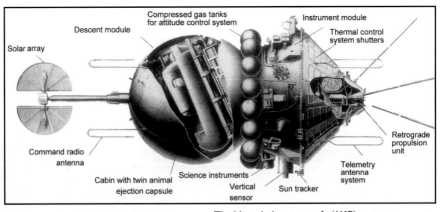

The Vostok-1 spacecraft (1KP).

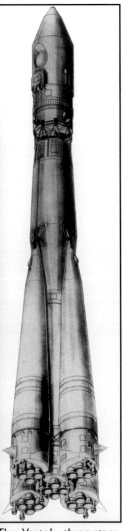

The Vostok three-stage launch vehicle consisted of a modified R-7 rocket and an E rocket unit with the spacecraft.

The Vostok spacecraft on the carriage in the shop.

Integration of the E rocket unit with the Vostok spacecraft.

Transportation of the Vostok launch vehicle with the Vostok (first manned) spacecraft to the launch complex.

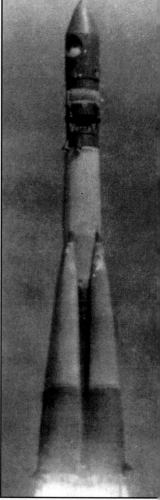

Yuri Alekseyevich Gagarin – Earth's first cosmonaut. Last steps on the ground before his historic flight.

The Vostok launch vehicle in flight

The Vostok launch vehicle engines firing.

The descent vehicle of the Vostok first manned spacecraft in the RSCE museum.

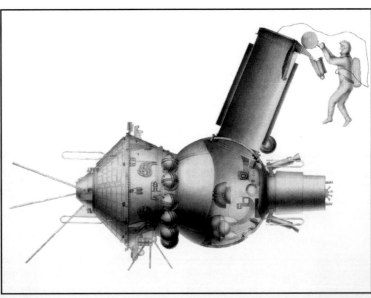

The Voskhod multi-man spacecraft made it possible to put a crew of three people into space, and as well provided a special airlock through which a man could egress into space.

Loading of the Voskhod spacecraft in he assembly-test building (ATB).

The airlock assembly of the Voskhod-2 spacecraft.

A.V. Keldysh inspects the Voshkod spacecraft.

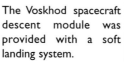

The Voskhod spacecraft descent module was provided with a soft landing system.

The Voskhod spacecraft on its support.

Preparing the Voskhod spacecraft for flight.

The Voskhod spacecraft as viewed from the BPS. The spacecraft has a back-up solid-propellant braking rocket engine.

Mating of the Voskhod spacecraft to the I rocket unit.

The Voskhod spacecraft payload shroud in the ATB.

Cosmonaut A. A. Leonov before flight. He was the first to egress into space and stayed there for 12 minutes and 9 seconds. He moved away from the spacecraft a distance of 5 meters.

Transportation of the launch vehicle with the Voskhod spacecraft to the launch pad.

Fueling of the launch vehicle for the Voskhod spacecraft.

Installation of the launch vehicle with the Voskhod spacecraft onto the launch pad.

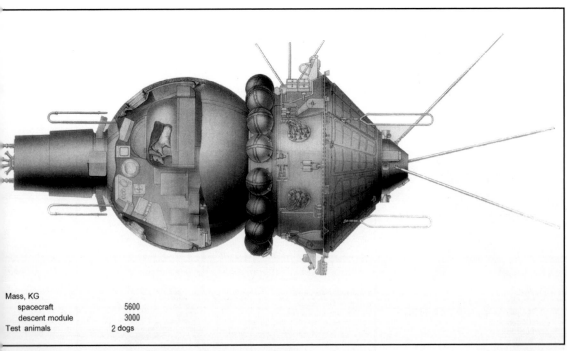

Mass, KG	
spacecraft	5600
descent module	3000
Test animals	2 dogs

The Voskhod research spacecraft, designed for long-term flight.

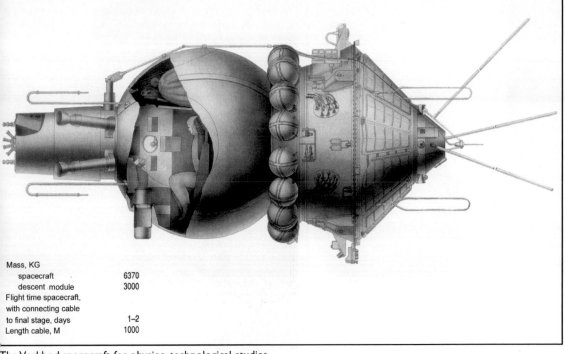

Mass, KG	
spacecraft	6370
descent module	3000
Flight time spacecraft, with connecting cable to final stage, days	1–2
Length cable, M	1000

The Voskhod spacecraft for physico-technological studies.

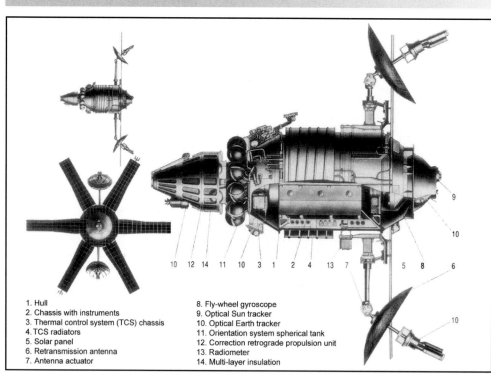

The Molniya satellite was the first communication satellite. It was put into a high-elliptic orbit and provided communication between the central regions and the far east.

1. Hull
2. Chassis with instruments
3. Thermal control system (TCS) chassis
4. TCS radiators
5. Solar panel
6. Retransmission antenna
7. Antenna actuator
8. Fly-wheel gyroscope
9. Optical Sun tracker
10. Optical Earth tracker
11. Orientation system spherical tank
12. Correction retrograde propulsion unit
13. Radiometer
14. Multi-layer insulation

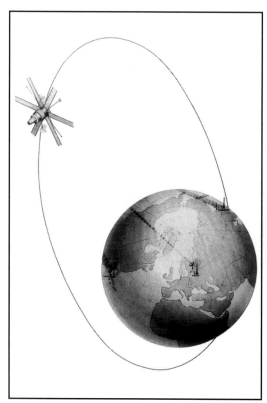

A twenty-four hour, long-range communication system was developed using the Molniya communication satellites.

The prototype of the Molniya communication satellite in the RSCE museum.

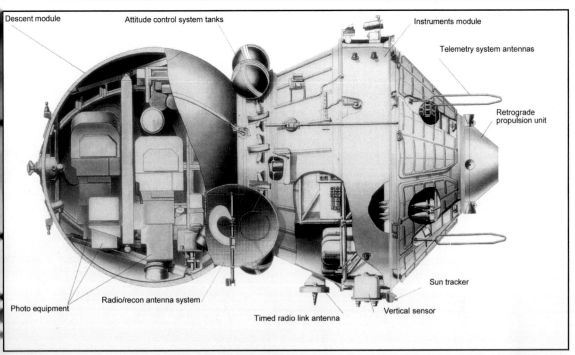

The Zenit-2 satellite. It was the first special-purpose unmanned satellite from which Earth photography was performed.

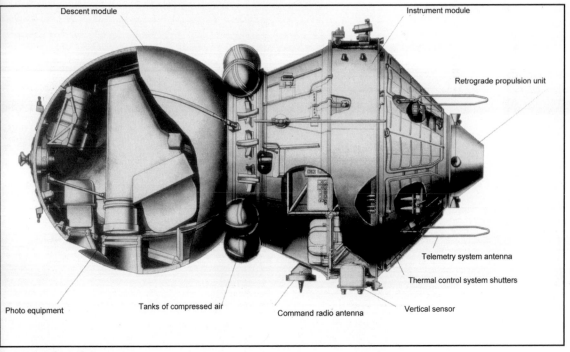

The Zenit-4 satellite.

The Zenit satellite, assembly and check before flight.

The Zenit satellite is prepared for mating with the rocket.

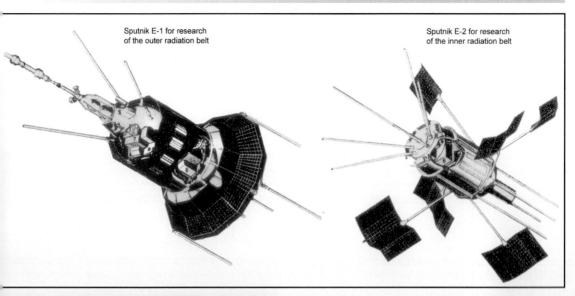

Sputnik E-1 for research of the outer radiation belt

Sputnik E-2 for research of the inner radiation belt

The Electron satellite system made it possible to get data on the radiation belt and the Earth's magnetic field that was necessary to provide radiation safety on manned flights.

Combat Missiles Designed in OKB-1

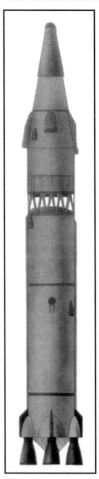

The R-9 oxygen-kerosene-fuelled missile.

The R-9 missile in flight. Its launch took place on April 9, 1961. In 1964 the missile complex was introduced into the inventory.

An R-9A missile near the Armed Forces Museum in Moscow.

The RT-2 missile (left) was the first intercontinental solid-propellant missile. Its first launch took place on February 26, 1966. In 1968 the missile was added to inventory.

The RT-1 missile (right). The first strategic solid-propellant missile. Its first launch took place on April 28, 1962.

The GR-1 three-stage global missile capable of destroying a target at any point on Earth from any direction.

Vasily Pavlovich Mishin
Chief designer of OKB-1
from 1966 until May 1974

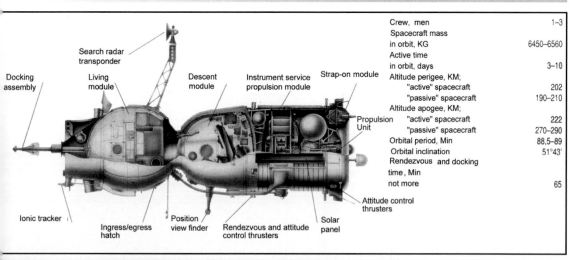

Crew, men		1–3
Spacecraft mass in orbit, KG		6450–6560
Active time in orbit, days		3–10
Altitude perigee, KM;	"active" spacecraft	202
	"passive" spacecraft	190–210
Altitude apogee, KM;	"active" spacecraft	222
	"passive" spacecraft	270–290
Orbital period, Min		88,5–89
Orbital inclination		51°43'
Rendezvous and docking time, Min not more		65

Labels on cutaway diagram: Docking assembly, Search radar transponder, Living module, Descent module, Instrument service propulsion module, Strap-on module, Propulsion Unit, Attitude control thrusters, Solar panel, Rendezvous and attitude control thrusters, Position view finder, Ingress/egress hatch, Ionic tracker

The Soyuz spacecraft (7K-OK) designed to execute a wide variety of tasks, including automatic and manual rendezvous, and docking with orbital spacecraft and stations.

The Soyuz spacecraft in the shop.

The Soyuz spacecraft on the mounting bogie.

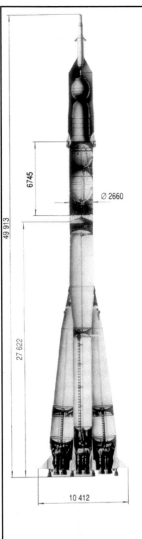

The Soyuz launch vehicle (11A511). The rocket houses the emergency crew recovery system which makes it possible to move the descent vehicle away from the rocket in distress.

The completion of the erection of the Soyuz launch vehicle with the spacecraft on the pad.

Mass fully fueled combined vehicle, T		315
Lift off mass combined vehicle, T		309
Mass lifted payload, T		6,8
Mass fuel component, T		277
Thrust engine, TF:		
Stage 1:		
on Earth		411
in vacuum		506
Stage 2 - in vacuum		100
Stage 3 - Module I		30,5
Specific thrust engine KGF-S/KG:		
Stage 1:		
on Earth		250
in vacuum		314
Stage 2:		
on Earth		254
in vacuum		311
Stage 3 - in vacuum		325

The Soyuz launch vehicle with the spacecraft ready for launch.

The Soyuz launch vehicle launch.

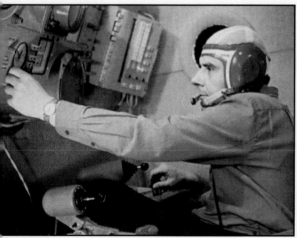

Cosmonaut V. M. Komarov operating the spacecraft rendezvous trainer.

The Soyuz launch vehicle ready for launch.

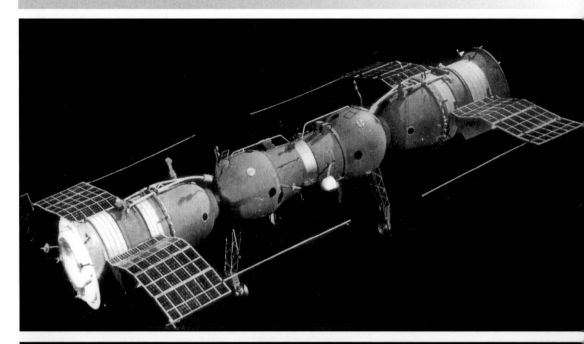

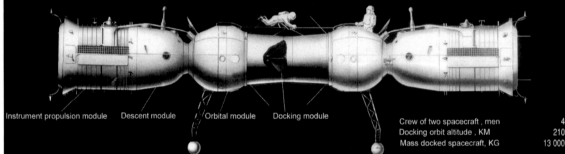

Instrument propulsion module   Descent module   Orbital module   Docking module

Crew of two spacecraft , men	4
Docking orbit altitude , KM	210
Mass docked spacecraft, KG	13 000

The Soyuz spacecraft (7K-OK) docking in orbit (top). The first docking of the Soyuz spacecraft was carried out in the automatic mode during the flight of the Cosmos-186 and Cosmos-188 unmanned spacecraft.

The Soyuz spacecraft docking in orbit. The Soyuz-4 and Soyuz-5 manned spacecraft docked on January 15, 1969. Cosmonauts transferred from one spacecraft to the other through space.

The Soyuz descent vehicle after landing.

The first flight to the Moon with return to the Earth

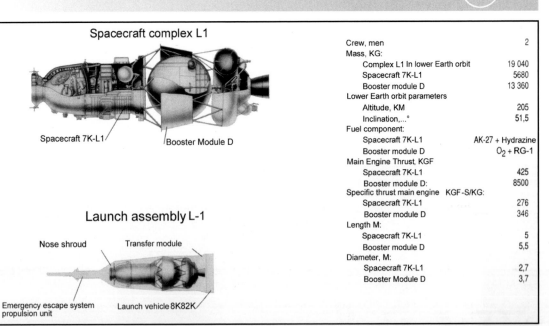

Spacecraft complex L1

Spacecraft 7K-L1

Booster Module D

Crew, men		2
Mass, KG:		
Complex L1 In lower Earth orbit		19 040
Spacecraft 7K-L1		5680
Booster module D		13 360
Lower Earth orbit parameters		
Altitude, KM		205
Inclination,...°		51,5
Fuel component:		
Spacecraft 7K-L1		AK-27 + Hydrazine
Booster module D		O_2 + RG-1
Main Engine Thrust, KGF		
Spacecraft 7K-L1		425
Booster module D:		8500
Specific thrust main engine KGF-S/KG:		
Spacecraft 7K-L1		276
Booster module D		346
Length M:		
Spacecraft 7K-L1		5
Booster module D		5,5
Diameter, M:		
Spacecraft 7K-L1		2,7
Booster Module D		3,7

Launch assembly L-1

Nose shroud

Transfer module

Emergency escape system propulsion unit

Launch vehicle 8K82K

The LI space complex for the circumlunar fly-by. This complex flew five times under the name of Zond. The complex flown used the D block rocket using multiple engine I I D58 firings

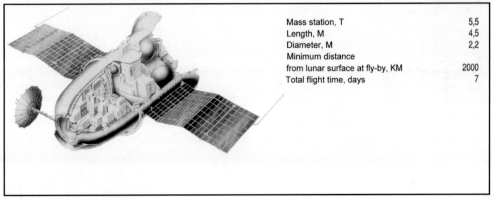

Mass station, T	5,5
Length, M	4,5
Diameter, M	2,2
Minimum distance	
from lunar surface at fly-by, KM	2000
Total flight time, days	7

The Zond automatic station.

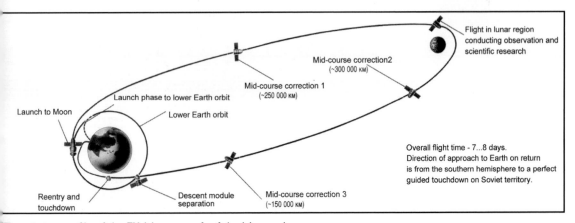

Flight in lunar region conducting observation and scientific research

Mid-course correction2 (~300 000 KM)

Mid-course correction 1 (~250 000 KM)

Launch phase to lower Earth orbit

Lower Earth orbit

Launch to Moon

Overall flight time - 7...8 days. Direction of approach to Earth on return is from the southern hemisphere to a perfect guided touchdown on Soviet territory.

Reentry and touchdown

Descent module separation

Mid-course correction 3 (~150 000 KM)

The mission profile of the 7K-LI spacecraft of the LI complex.

Transportation of the LI complex to the launch area.

The Proton launch vehicle with the LI complex on the launch pad.

Launch of the Proton vehicle with the LI complex.

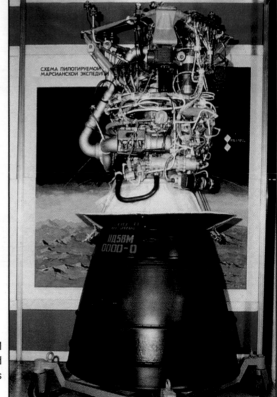

The close-cycle liquid fuel rocket engine 11D58M of TsKBEM development. This engine of 8.5 tons thrust uses oxygen and hydrocarbon fuel as propellant components. It was the world's first engine to provide multiple in-flight firing.

The Earth and Moon photographs taken with photographic equipment on board the Zond-5 and Zond-6 stations.

The descent vehicle of the Zond-5 station in the Indian Ocean after its return from circumlunar flight.

Lunar Manned Program activities

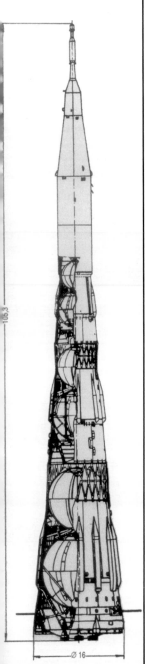

In accordance with the lunar manned program, the N1-L3 system was designed, which included the N1 three-stage rocket and L3 lunar complex.

The A block was used as the first stage of the N1 rocket.

The maximum diameter of the block is 16.8 meters (dimensions taken by stabilizers are 22.33 meters) with a height of 30.1 meters. The block houses 30 engines with ground thrust of 153 tons each.

Payload mass capacity to low Earth orbit (Orbital altitude = 200 км), т	90
Lift off Mass, T	2800
Mass fuel, T:	
Oxygen	1730
Kerosene	680
Total thrust engines on Earth, TF	4615
Total length, M	105,3

The V block was used as the third stage. Maximum diameter of the block is about 7.6 meters with the height (by the interfaces) being 11.5 meters. The block houses four engines with a vacuum thrust of 41 tons each.

The B block was used as the second stage of the N1 rocket. The maximum diameter of the block is about 10.3 meters with a height of 20.5 meters. The block houses 8 engines with a vacuum thrust of 180 tons each.

The N1 launch vehicle on the mounting bogie in the assembly-test building of the cosmodrome.

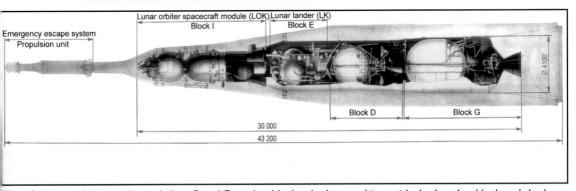

The L3 lunar rocket complex including G and D rocket blocks, the lunar orbiter with the I rocket block and the lunar vehicle with E rocket block.

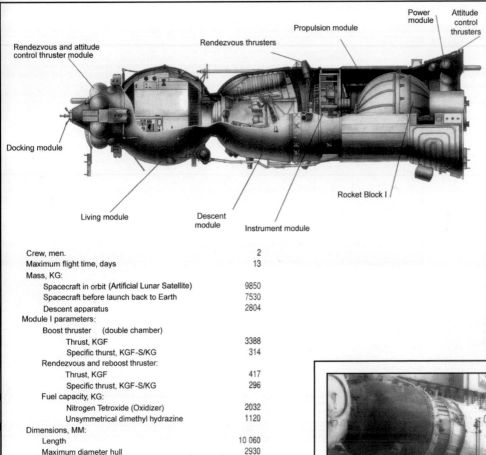

Crew, men.	2
Maximum flight time, days	13
Mass, KG:	
Spacecraft in orbit (Artificial Lunar Satellite)	9850
Spacecraft before launch back to Earth	7530
Descent apparatus	2804
Module I parameters:	
Boost thruster (double chamber)	
Thrust, KGF	3388
Specific thurst, KGF-S/KG	314
Rendezvous and reboost thruster:	
Thrust, KGF	417
Specific thrust, KGF-S/KG	296
Fuel capacity, KG:	
Nitrogen Tetroxide (Oxidizer)	2032
Unsymmetrical dimethyl hydrazine	1120
Dimensions, MM:	
Length	10 060
Maximum diameter hull	2930

The lunar orbiter including the habitation compartment and the vehicle to be descended to Earth, as well as the I rocket unit, and the instrumentation and service module. The orbiter mass in ALS orbit is 9,850 kg. Maximum length is about 10 meters, diameter being 2.9 meters.

The lunar orbiter on the mounting bogie.

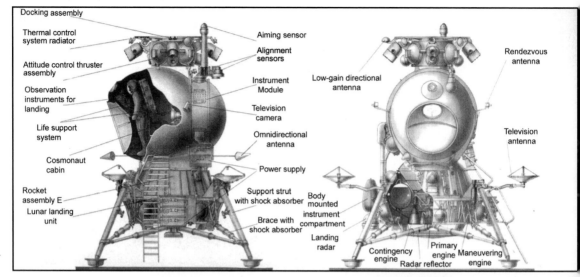

Docking assembly

Thermal control system radiator

Attitude control thruster assembly

Observation instruments for landing

Life support system

Cosmonaut cabin

Rocket assembly E

Lunar landing unit

Aiming sensor

Alignment sensors

Instrument Module

Television camera

Omnidirectional antenna

Power supply

Support strut with shock absorber

Brace with shock absorber

Low-gain directional antenna

Rendezvous antenna

Television antenna

Body mounted instrument compartment

Landing radar

Contingency engine

Radar reflector

Primary engine

Maneuvering engine

The lunar vehicle consisting of the lunar descent assembly, the cosmonaut's cabin with various systems, and the E rocket unit with main and stand-by engines.

The lunar vehicle in the shop.

The 7K-L1S unmanned space vehicle used during the first launch of the N1 rocket, instead of the orbiter and the lunar vehicle, on February 21, 1969.

The L3 lunar complex in the assembly-test building.

The 7K-L1S unmanned spacecraft in the assembly jig.

The N1-L3 system on the way to the launching complex.

The N1-L3 system near the launch pad.

The N1-L3 space system on the transport-erecting assembly in the assembly building, ready for roll-out to the launch area.

The N1-L3 system is erected vertically on the launcher. The transport-erecting assembly is not moved away.

Erection of the N1-L3 system on the launcher.

The N1-L3 system on the launcher ready for launch.

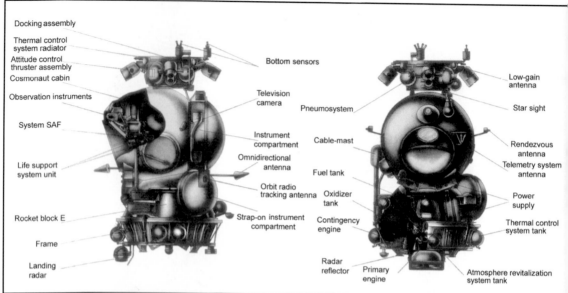

Docking assembly
Thermal control system radiator
Attitude control thruster assembly
Cosmonaut cabin
Observation instruments
System SAF
Life support system unit
Rocket block E
Frame
Landing radar

Bottom sensors
Television camera
Pneumosystem
Instrument compartment
Cable-mast
Omnidirectional antenna
Fuel tank
Orbit radio tracking antenna
Oxidizer tank
Strap-on instrument compartment
Contingency engine
Radar reflector
Primary engine

Low-gain antenna
Star sight
Rendezvous antenna
Telemetry system antenna
Power supply
Thermal control system tank
Atmosphere revitalization system tank

The T2K space vehicle was used for developing the lunar vehicle's systems under space conditions in near-Earth orbit.

The development of the lunar vehicle landing on a special mock-up.

The T2K space vehicle launch into orbit.

Launch of the N1-L3 system.

A view of the Soyuz spacecraft from the
Apollo spacecraft.

The crew on board the docked Soyuz and
Apollo spacecraft.

The androgynous peripheral docking assembly designed for the Soyuz and Apollo docking.

The Soyuz-19 descent vehicle in the RSCE museum.

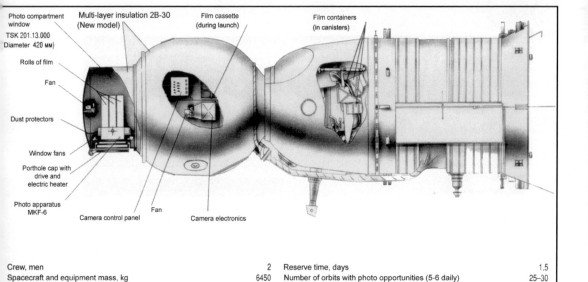

Photo compartment window
TSK 201.13.000
Diameter 420 мм)

Multi-layer insulation 2B-30 (New model)

Film cassette (during launch)

Film containers (in canisters)

Rolls of film

Fan

Dust protectors

Window fans

Porthole cap with drive and electric heater

Photo apparatus MKF-6

Camera control panel

Fan

Camera electronics

Crew, men	2	Reserve time, days	1.5
Spacecraft and equipment mass, kg	6450	Number of orbits with photo opportunities (5-6 daily)	25–30
Orbital altitude, KM	~260	Landing orbits (daily)	1-й, 6-й
Orbital inclination,... °	≈65	Photo session duration (per revolution), no less than	16
Latitudinal range of photography, ...° North latitude	35–65	Soviet territory coverage capability, %	20–30
Width of photographed area (along vehicle translation path) KM	≈150		
Flight time (nominal), days	6		

The Soyuz-22 spacecraft (which is modified from the back-up spacecraft in ASTP) to be launched as part of the intercosmos program, was equipped with the MKF-6 multizone photographic camera to test the methods and means of studying geological and geographical characteristics of the Earth's surface from space for the benefit of national economy and environmental control.

Development of Orbital Stations

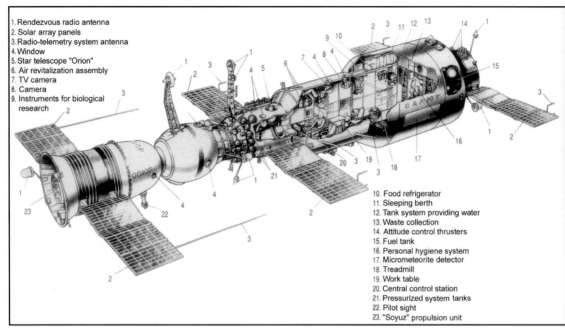

1. Rendezvous radio antenna
2. Solar array panels
3. Radio-telemetry system antenna
4. Window
5. Star telescope "Orion"
6. Air revitalization assembly
7. TV camera
8. Camera
9. Instruments for biological
   research

10. Food refrigerator
11. Sleeping berth
12. Tank system providing water
13. Waste collection
14. Attitude control thrusters
15. Fuel tank
16. Personal hygiene system
17. Micrometeorite detector
18. Treadmill
19. Work table
20. Central control station
21. Pressurized system tanks
22. Pilot sight
23. "Soyuz" propulsion unit

The Soyuz transport spacecraft (of 7K-T type) docked to the Salyut station.

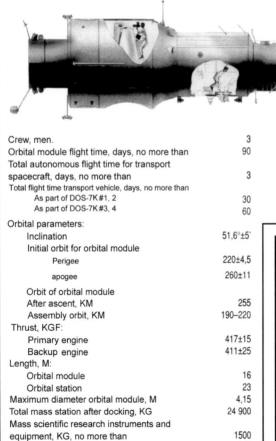

The first long-term orbital station (DOS-7K had only one docking assembly.

Crew, men.	3
Orbital module flight time, days, no more than	90
Total autonomous flight time for transport spacecraft, days, no more than	3
Total flight time transport vehicle, days, no more than	
As part of DOS-7K #1, 2	30
As part of DOS-7K #3, 4	60
Orbital parameters:	
Inclination	$51,6°±5'$
Initial orbit for orbital module	
Perigee	220±4,5
apogee	260±11
Orbit of orbital module	
After ascent, KM	255
Assembly orbit, KM	190–220
Thrust, KGF:	
Primary engine	417±15
Backup engine	411±25
Length, M:	
Orbital module	16
Orbital station	23
Maximum diameter orbital module, M	4,15
Total mass station after docking, KG	24 900
Mass scientific research instruments and equipment, KG, no more than	1500

The Salyut orbital station on the mounting bogie.

Preparation of the orbital station for mating with the Proton launch vehicle.

The Salyut-2 orbital station, known as Cosmos-557, in the assembly-test building.

The Proton launch vehicle with the first Zarya orbital station, which was called Salyut in the press, on the launch pad.

The Salyut-4 orbital station on the mounting bogie.

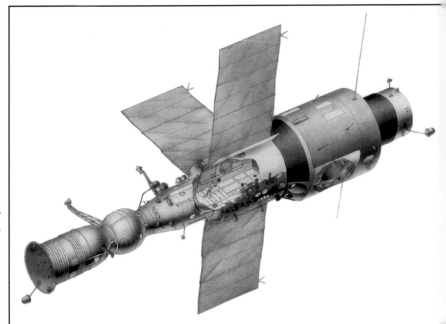

The Salyut-4 orbital station.

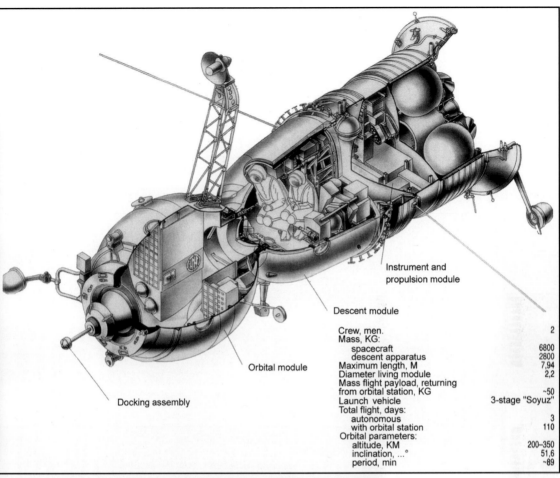

Instrument and
propulsion module

Descent module

Orbital module

Docking assembly

Crew, men.	2
Mass, KG:	
spacecraft	6800
descent apparatus	2800
Maximum length, M	7,94
Diameter living module	2,2
Mass flight payload, returning from orbital station, KG	~50
Launch vehicle	3-stage "Soyuz"
Total flight, days:	
autonomous	3
with orbital station	110
Orbital parameters:	
altitude, KM	200–350
inclination, ...°	51,6
period, min	~89

The Soyuz-type transport space-craft (7K-T) used to supply the first generation orbital stations.

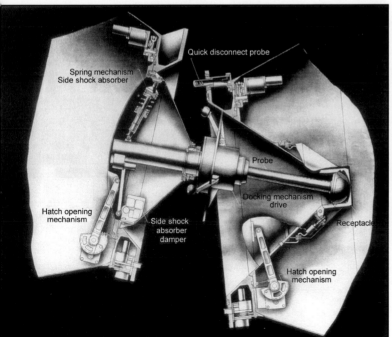

Quick disconnect probe

Spring mechanism
Side shock absorber

Probe

Hatch opening
mechanism

Docking mechanism
drive

Side shock
absorber
damper

Receptacle

Hatch opening
mechanism

The Soyuz-10 spacecraft (7K-T) and subsequent craft had docking assemblies with a central transfer hatch through which cosmonauts could transfer from one spacecraft to the other without egressing into space.

The Soyuz type spacecraft in its assembly jig.

Emergency Descent System Operating Sequence During Ascent

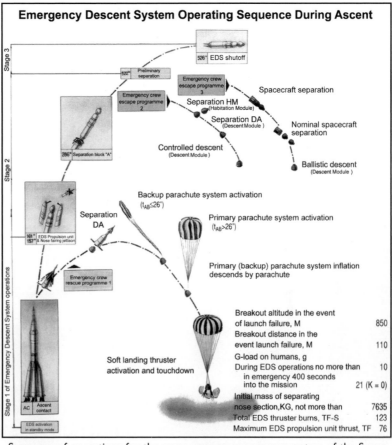

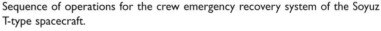

The Soyuz descent vehicle landing. The descent vehicle was about 3 tons in mass, and 2.2 meters in diameter (over thermal protection). Its configuration is similar to a "head-light" (the lift-to-drag ratio was 0.30).

Breakout altitude in the event of launch failure, M	850
Breakout distance in the event launch failure, M	110
G-load on humans, g	
During EDS operations no more than	10
in emergency 400 seconds into the mission	21 (K = 0)
Initial mass of separating nose section, KG, not more than	7635
Total EDS thruster burns, TF-S	123
Maximum EDS propulsion unit thrust, TF	76

Sequence of operations for the crew emergency recovery system of the Soyuz T-type spacecraft.

The Soyuz T spacecraft descent vehicle at the landing site.

Operation of
the Soyuz
spacecraft
landing
complex.

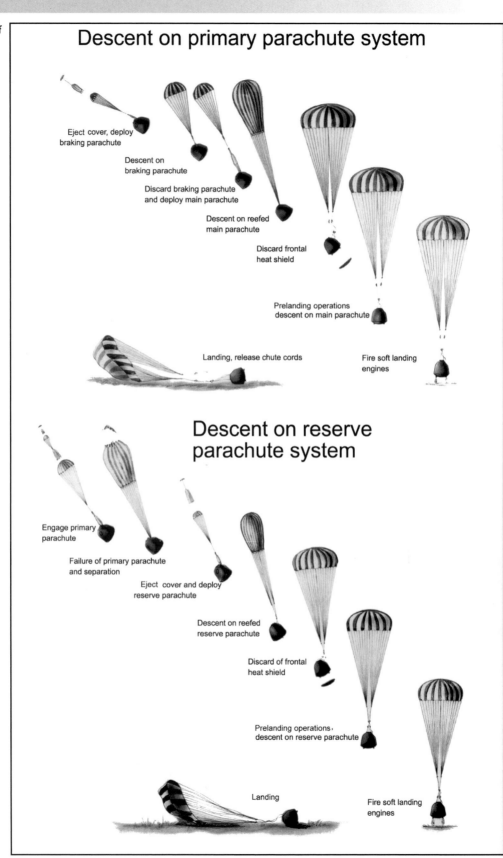

Descent on primary parachute system

Eject cover, deploy
braking parachute

Descent on
braking parachute

Discard braking parachute
and deploy main parachute

Descent on reefed
main parachute

Discard frontal
heat shield

Prelanding operations
descent on main parachute

Landing, release chute cords

Fire soft landing
engines

Descent on reserve parachute system

Engage primary
parachute

Failure of primary parachute
and separation

Eject cover and deploy
reserve parachute

Descent on reefed
reserve parachute

Discard of frontal
heat shield

Prelanding operations,
descent on reserve parachute

Landing

Fire soft landing
engines

Stages of the Salyut orbital station development

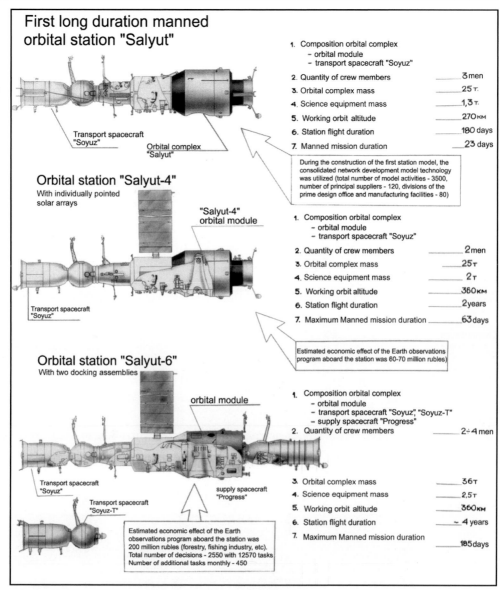

First long duration manned orbital station "Salyut"

Transport spacecraft "Soyuz"

Orbital complex "Salyut"

1. Composition orbital complex
   - orbital module
   - transport spacecraft "Soyuz"
2. Quantity of crew members _____ 3 men
3. Orbital complex mass _____ 25 т.
4. Science equipment mass _____ 1,3 т.
5. Working orbit altitude _____ 270 км
6. Station flight duration _____ 180 days
7. Manned mission duration _____ 23 days

During the construction of the first station model, the consolidated network development model technology was utilized (total number of model activities - 3500, number of principal suppliers - 120, divisions of the prime design office and manufacturing facilities - 80)

Orbital station "Salyut-4"
With individually pointed solar arrays

"Salyut-4" orbital module

Transport spacecraft "Soyuz"

1. Composition orbital complex
   - orbital module
   - transport spacecraft "Soyuz"
2. Quantity of crew members _____ 2 men
3. Orbital complex mass _____ 25 т
4. Science equipment mass _____ 2 т
5. Working orbit altitude _____ 350 км
6. Station flight duration _____ 2 years
7. Maximum Manned mission duration _____ 63 days

Estimated economic effect of the Earth observations program aboard the station was 60-70 million rubles)

Orbital station "Salyut-6"
With two docking assemblies

orbital module

Transport spacecraft "Soyuz"

Transport spacecraft "Soyuz-T"

supply spacecraft "Progress"

Estimated economic effect of the Earth observations program aboard the station was 200 million rubles (forestry, fishing industry, etc). Total number of decisions - 2550 with 12570 tasks Number of additional tasks monthly - 450

1. Composition orbital complex
   - orbital module
   - transport spacecraft "Soyuz", "Soyuz-T"
   - supply spacecraft "Progress"
2. Quantity of crew members _____ 2÷4 men
3. Orbital complex mass _____ 36 т
4. Science equipment mass _____ 2,5 т
5. Working orbit altitude _____ 350 км
6. Station flight duration _____ ~ 4 years
7. Maximum Manned mission duration _____ 185 days

The Salyut-6 orbital station (second generation station) had two docking assemblies to which the transport and cargo spacecraft could be docked.

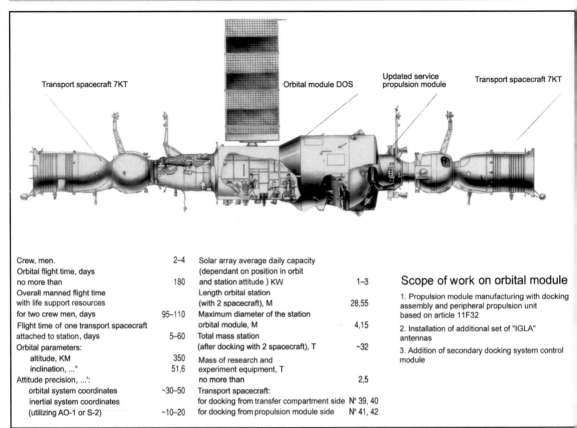

Transport spacecraft 7KT   Orbital module DOS   Updated service   Transport spacecraft 7KT
                                                         propulsion module

		Scope of work on orbital module
Crew, men.	2–4	
Orbital flight time, days no more than	180	Solar array average daily capacity (dependant on position in orbit and station attitude) KW 1–3
Overall manned flight time with life support resources for two crew men, days	95–110	Length orbital station (with 2 spacecraft), M 28,55
Flight time of one transport spacecraft attached to station, days	5–60	Maximum diameter of the station orbital module, M 4,15
Orbital parameters:		Total mass station (after docking with 2 spacecraft), T ~32
altitude, KM	350	
inclination, ...°	51,6	Mass of research and experiment equipment, T
Attitude precision, ...':		no more than 2,5
orbital system coordinates	~30–50	Transport spacecraft:
inertial system coordinates (utilizing AO-1 or S-2)	~10–20	for docking from transfer compartment side № 39, 40
		for docking from propulsion module side № 41, 42

Scope of work on orbital module

1. Propulsion module manufacturing with docking assembly and peripheral propulsion unit based on article 11F32

2. Installation of additional set of "IGLA" antennas

3. Addition of secondary docking system control module

The Salyut-6 orbital station (second generation station) had two docking assemblies to which the transport and cargo spacecraft could be docked.

The Salyut-6 orbital station with the docked Soyuz T spacecraft in flight..

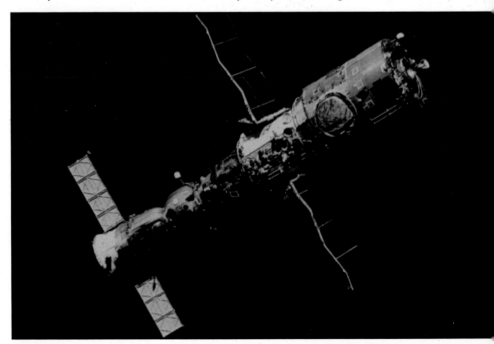

The Salyut-7 orbital station with the docked Soyuz T spacecraft in flight.

The Salyut-7 orbital station with the Cosmos-686 cargo spacecraft in automatic flight mode.

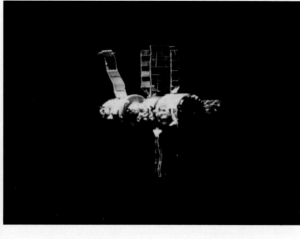

The Soyuz T transport spacecraft with solar arrays.

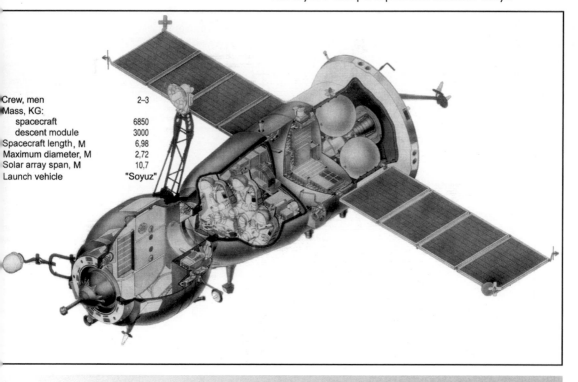

Crew, men	2–3
Mass, KG:	
spacecraft	6850
descent module	3000
Spacecraft length, M	6,98
Maximum diameter, M	2,72
Solar array span, M	10,7
Launch vehicle	"Soyuz"

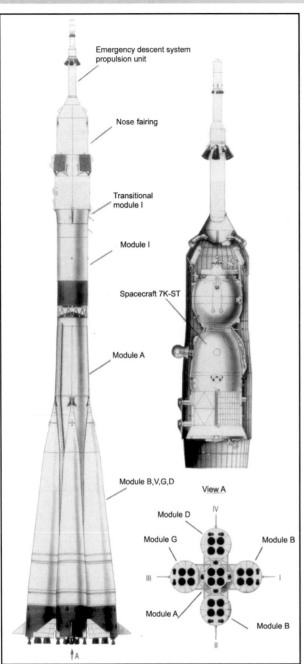

Emergency descent system propulsion unit

Nose fairing

Transitional module I

Module I

Spacecraft 7K-ST

Module A

Module B,V,G,D

View A

Module D

Module G   Module B

Module A

Module B

The 11A511Y launch vehicle with the 7K-ST (Soyuz T) spacecraft.

Launch vehicle	11A511У		11A511У-2
Initial orbital parameters:			
inclination, ...°	51,6		64,8
altitude (average), KM		220	
Lift-off mass, T:			
carrier stage and nose fairing with payload	309,7		310,0
spacecraft	6,855		6,740
Crew, men	2–3		2–3
Number stages on launch vehicle		3	
Stage 1		module A,B,V,G,D	
Stage 2		module A	
Stage 3		module I	
Fuel component			
module A		oxygen + kerosene	oxygen + cycline
module B,V,G,D,I		oxygen + kerosene	
Maximum engine thrust TC			
Stage 1			
on Earth	413,3		420,0
in vacuum	505,3		509,5
Stage 2 - in vacuum	99,7		103,1
Stage 3 - in vacuum		30,4	
Length of launch vehicle and			
nose fairing with payload	51,1		51,3
Maximum transverse dimensions, M		10,3	

Transportation of the Soyuz launch vehicle with the Soyuz spacecraft to the launching area.

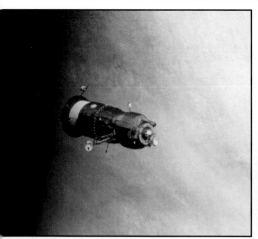

The Progress cargo spacecraft in flight.

The launch vehicle with the Progress spacecraft.

The Progress cargo spacecraft.

Mass spacecraft, KM	7020
Delivered payload weight, KG	~2300
including:	
in cargo module, KG, not more than	1300
in propellant module, KG	
not more than	1000
Maximum length, M	7,94
Maximum diameter	
pressurized module, M	2,2
Launch vehicle	"Soyuz"
Flight time , days, not more than:	
autonomous	3
with orbital station	30
Orbital parameters	
altitude, KM	200–350
inclination, ...°	51,6
period, min	~89

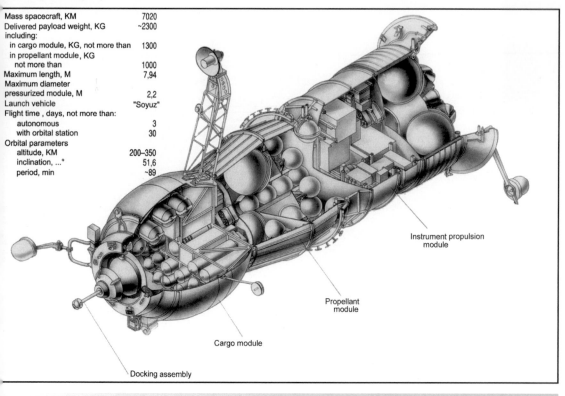

Instrument propulsion module

Propellant module

Cargo module

Docking assembly

Yuri Pavlovich Semenov
General Director and General Designer
of S. P. Korolev NPO Energia since 1989

The first Permanently Operating Mir Complex in Orbit

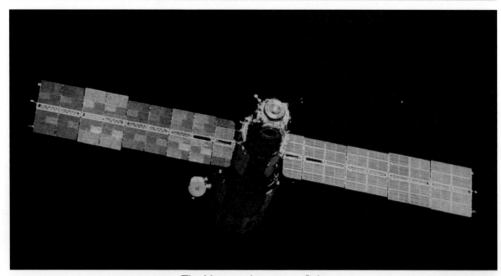

The Mir complex core in flight.

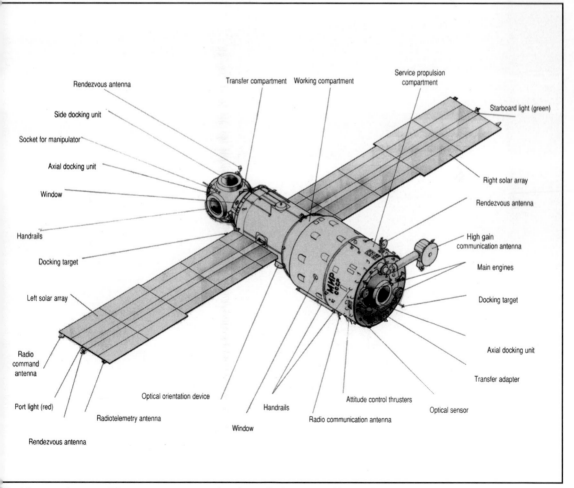

Rendezvous antenna

Side docking unit

Socket for manipulator

Axial docking unit

Window

Handrails

Docking target

Left solar array

Radio command antenna

Port light (red)

Rendezvous antenna

Transfer compartment Working compartment

Service propulsion compartment

Starboard light (green)

Right solar array

Rendezvous antenna

High gain communication antenna

Main engines

Docking target

Axial docking unit

Transfer adapter

Optical orientation device

Radiotelemetry antenna

Window

Handrails

Radio communication antenna

Attitude control thrusters

Optical sensor

The Mir complex core. The core was equipped with six docking assemblies and a new docking system.

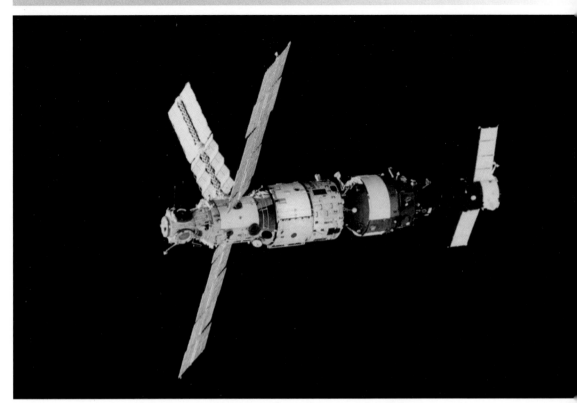

The Mir complex core with the Kvant module and the Soyuz TM spacecraft in flight.

The launch of the Mir complex core. The launch was accomplished with the use of the Proton launch vehicle on February 20, 1986.

The Soyuz launch vehicle flight.

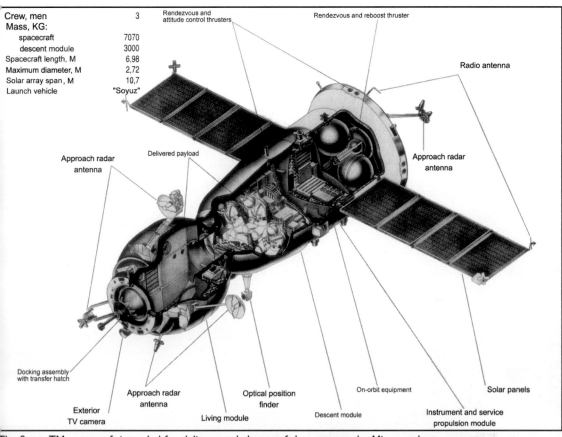

Crew, men	3
Mass, KG:	
spacecraft	7070
descent module	3000
Spacecraft length, M	6,98
Maximum diameter, M	2,72
Solar array span, M	10,7
Launch vehicle	"Soyuz"

Rendezvous and attitude control thrusters

Rendezvous and reboost thruster

Radio antenna

Approach radar antenna

Delivered payload

Approach radar antenna

Approach radar antenna

Docking assembly with transfer hatch

Exterior TV camera

Approach radar antenna

Living module

Optical position finder

Descent module

On-orbit equipment

Solar panels

Instrument and service propulsion module

The Soyuz TM spacecraft intended for delivery and change of the crew on the Mir complex.

Installation of the Soyuz launch vehicle with Soyuz TM spacecraft onto the launching pad.

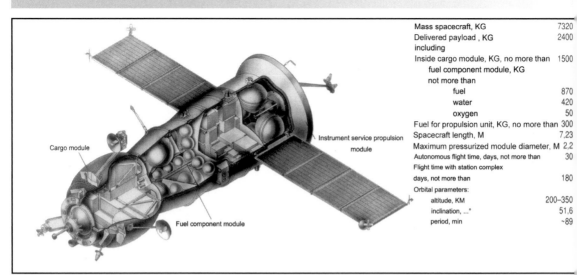

Mass spacecraft, KG	7320
Delivered payload , KG including	2400
Inside cargo module, KG, no more than	1500
fuel component module, KG not more than	
fuel	870
water	420
oxygen	50
Fuel for propulsion unit, KG, no more than	300
Spacecraft length, M	7,23
Maximum pressurized module diameter, M	2,2
Autonomous flight time, days, not more than	30
Flight time with station complex days, not more than	180
Orbital parameters:	
altitude, KM	200–350
inclination, ...°	51,6
period, min	~89

The Progress M cargo spacecraft intended for delivery of fuel and other consumables to the Mir complex.

An international crew on board the Mir complex.

Launch of the Soyuz launch vehicle with the Progress M cargo spacecraft.

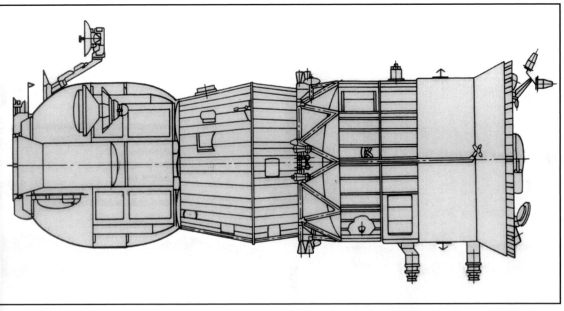

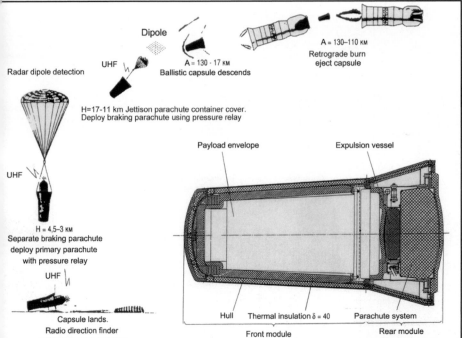

Dipole

UHF

Radar dipole detection

A = 130 - 17 км
Ballistic capsule descends

A = 130–110 км
Retrograde burn
eject capsule

H=17-11 km Jettison parachute container cover.
Deploy braking parachute using pressure relay

UHF

H = 4,5–3 км
Separate braking parachute
deploy primary parachute
with pressure relay

UHF

Payload envelope

Expulsion vessel

Capsule lands.
Radio direction finder
and light beacon activates

Hull   Thermal insulation δ = 40   Parachute system

Front module

Rear module

The Progress M spacecraft equipped with the recovery ballistic capsule.

The recovery ballistic capsule made it possible to deliver results of investigations carried out by cosmonauts on board the Mir complex back to Earth.

The recovery ballistic capsule with the parachute in the RSCE museum.

		Objectives
Mass capsule (max), KG	350	Quick return of self-financing
Mass returning payload, KG, no more than	150	projects and commercial
Cargo spacecraft retrograde burn (ΔVт), M/S	~150	contracts in the area
Velocity of descent on principal parachute, M/S	8	of technology, biology,
Touchdown precision		photography (films, magnetic
(ΔVт ~150 M/S, H_{orb} ~350 KM:		tapes, kits with experiment
along the route	±125	results).
lateral spread	±15	
Time to detection by base, hrs	3	
Expected flow retrieved cargo		
by rapid retrieval 2-3 capsules per year		
(1991–1994), KG, not more than	1200	
Payload recovery opportunities on 11F732 spacecraft		
(12 spacecraft), KG, not more than	600	

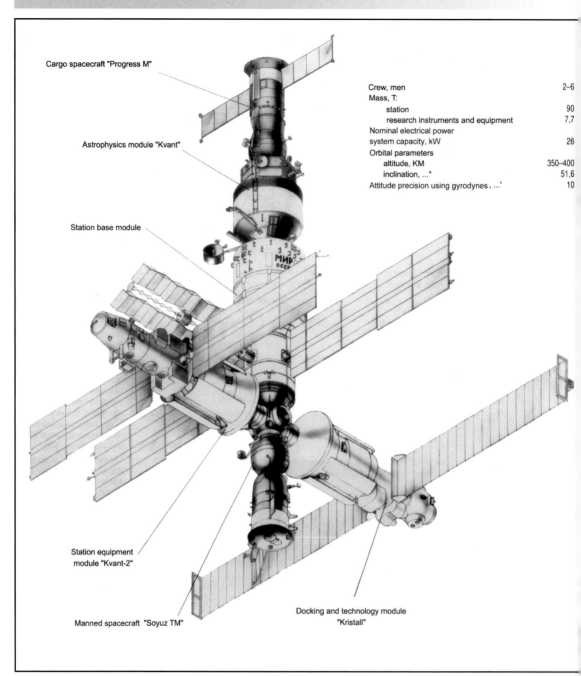

Cargo spacecraft "Progress M"

Astrophysics module "Kvant"

Station base module

Station equipment
module "Kvant-2"

Manned spacecraft "Soyuz TM"

Docking and technology module
"Kristall"

Crew, men	2–6
Mass, T:	
station	90
research instruments and equipment	7,7
Nominal electrical power	
system capacity, kW	26
Orbital parameters	
altitude, KM	350–400
inclination, ...°	51,6
Attitude precision using gyrodynes , ...'	10

The arrangement of modules on the Mir orbital complex core.

The Mir complex with the Kvant, Kvant-2 and Kristall modules, transport spacecraft Soyuz TM-16, cargo spacecraft Progress M-17 and undocked cargo spacecraft Progress M-18. The picture was taken from the Soyuz TM-17 transport spacecraft on July 3, 1993.

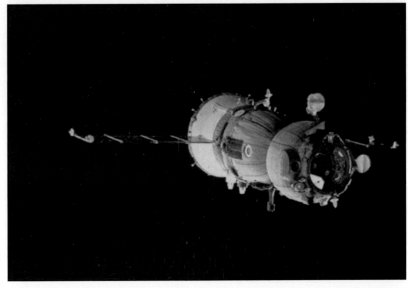

The Soyuz TM-16 spacecraft in flight.

Development of the Energia Launch Vehicle and Buran orbiter

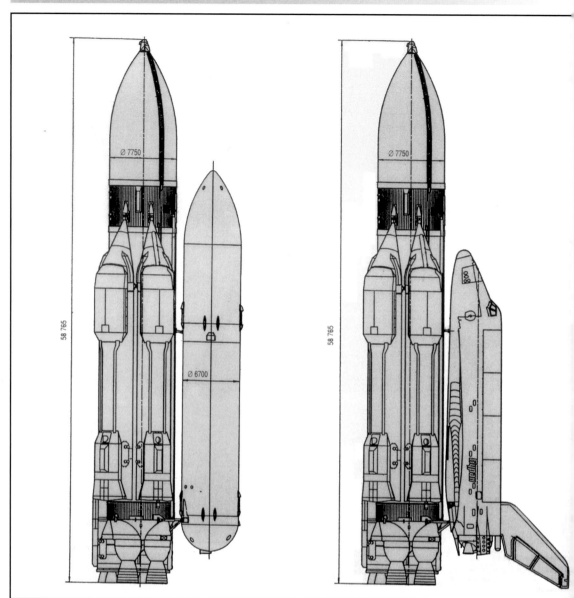

The Energia launch vehicle. The first stage consists of four side modules, the second stage is the central module. Engines of all modules fire at the moment of ignition. The payload is fastened to the side of the central module. For a payload, the Energia launch vehicle can have the Buran orbiter or the cargo transport container (6.7 meters in diameter where large-scale load and the booster unit are located).

Preparation of the side modules for assembly of the launch vehicle. The RD170 four-chamber engine (740 tons thrust near the ground; 806 tons in vacuum) is mounted on the module.

Assembly of the Energia launch vehicle is performed in the assembly-test building of the cosmodrome. The picture shows the span of the building with the first stage modules and the assembled rocket "package".

The versatile stand-start complex for performing firing tests of the launch vehicle and for launching.

The Energia launch vehicle at the versatile stand-start complex.

Transportation of the Energia launch vehicle is accomplished with the use of a special transport-erecting assembly.

The Energia launch vehicle at the launching complex.

Transportation of the Energia launch vehicle (with the Polus spacecraft on the external suspension) to the versatile stand-start complex.

The Energia launch vehicle with the Polus spacecraft on the versatile stand-start complex being prepared for its first launch.

The first launch of the Energia launch vehicle took place at 21:30 Moscow time on May 15, 1987.

The Buran analog was equipped with four engines permitting its take-off from the aerodrome strip. This allowed it to be used for the testing and development of orbiter piloting operations to be used during landing following orbital flight.

Mating the Buran to the Energia launch vehicle.

Installing the Energia – Buran system onto the transport-erecting assembly.

Transportation of the Energia – Buran system to the launching complex.

The Energia – Buran system erected on the launcher. The lifting device of the transport-erecting assembly is now vertical.

The Energia – Buran system on the launch pad.

The first launch of the Energia – Buran system took place at 6:00 Moscow time on November 15, 1988.

The Buran approach and landing on the cosmodrome's airfield runway after two-orbit orbital flight.

Trend of development

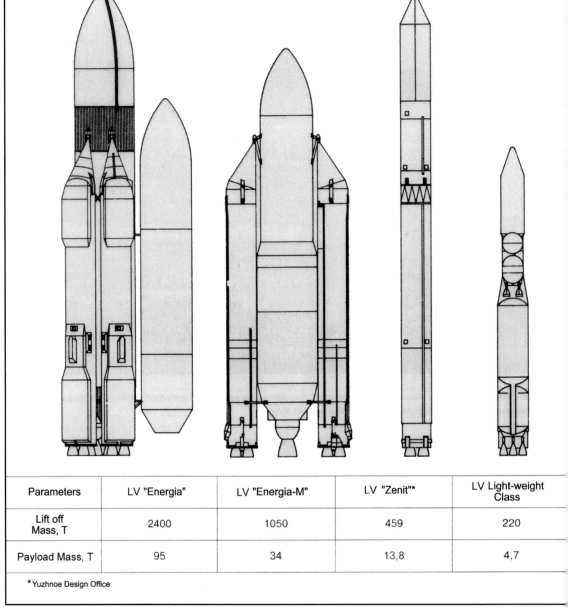

Parameters	LV "Energia"	LV "Energia-M"	LV "Zenit"*	LV Light-weight Class
Lift off Mass, T	2400	1050	459	220
Payload Mass, T	95	34	13,8	4,7

*Yuzhnoe Design Office

A number of launch vehicles designed on the basis of the Energia launch vehicle. They use the same elements (modules, engines, etc.), which substantially reduces the time frame of their development.

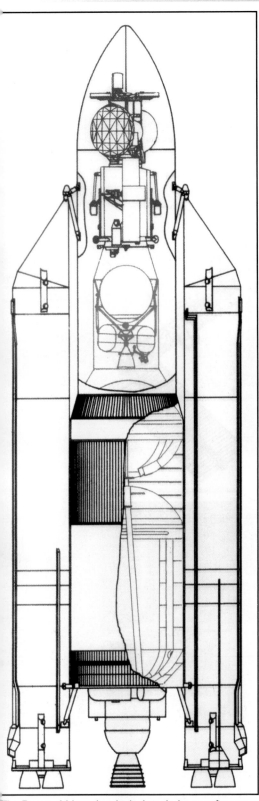

Transportation of the Energia-M launch vehicle is accomplished on the transport-erecting assembly of the Energia launch vehicle.

The Energia-M launch vehicle erected on the versatile stand-start complex.

The Energia-M launch vehicle. It includes two first stage Energia side modules and a shortened central module with one engine. The payload is located under the nose fairing above the central module.

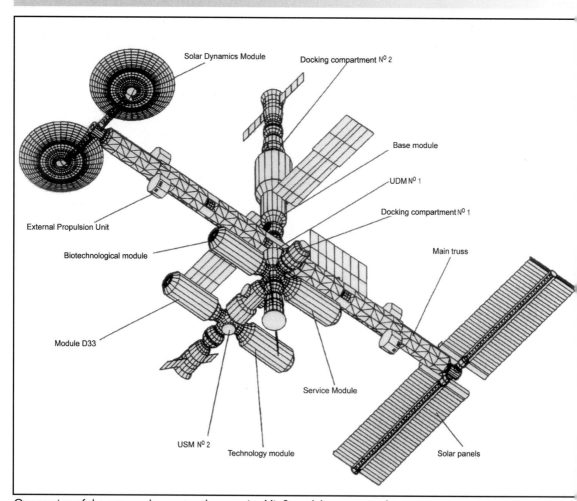

One version of the proposed permanently operating Mir-2 modular-type complex.

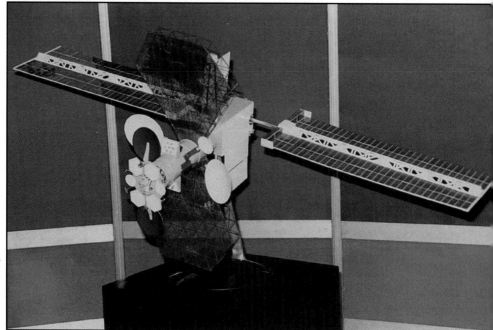

A mock-up of the versatile space platform.

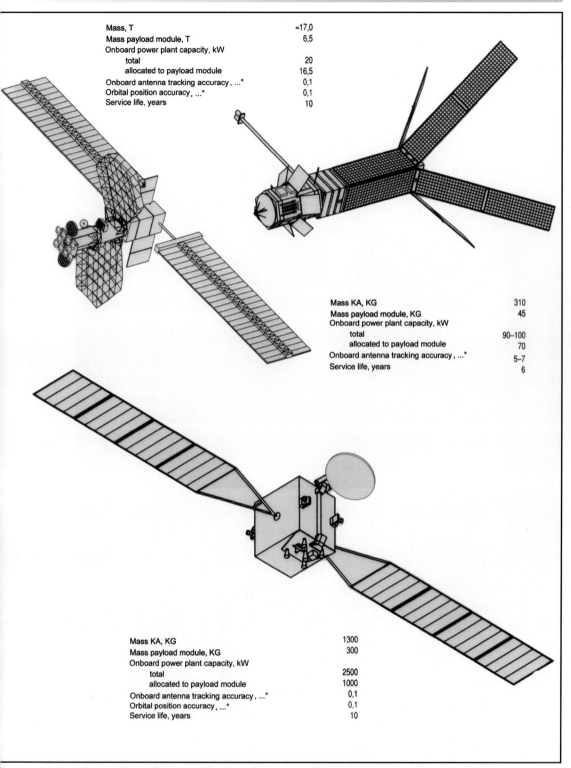

Mass, T	≈17,0
Mass payload module, T	6,5
Onboard power plant capacity, kW	
total	20
allocated to payload module	16,5
Onboard antenna tracking accuracy, ...°	0,1
Orbital position accuracy, ...°	0,1
Service life, years	10

Mass KA, KG	310
Mass payload module, KG	45
Onboard power plant capacity, kW	
total	90–100
allocated to payload module	70
Onboard antenna tracking accuracy, ...°	5–7
Service life, years	6

Mass KA, KG	1300
Mass payload module, KG	300
Onboard power plant capacity, kW	
total	2500
allocated to payload module	1000
Onboard antenna tracking accuracy, ...°	0,1
Orbital position accuracy, ...°	0,1
Service life, years	10

Proposed designs for the Globis, Signal and Yamal satellite communication system components.

Artist's concept of the US Space Shuttle docked to the Mir orbital station.

Artist's concept of the completed International space station.

The Yamal 102 communications satellite. The first Yamal satellite was launched on September 6, 1999. The Yamal satellites were built for AO Gazcom of Moscow, a joint venture of Energia and RAO Gazprom, the Russian natural gas company.

The Yamal satellites have a communications payload of 12 C-band transponders (built by Space Systems/Loral) and are equipped with Fakel SPD-70 plasma thrusters for inclination control.

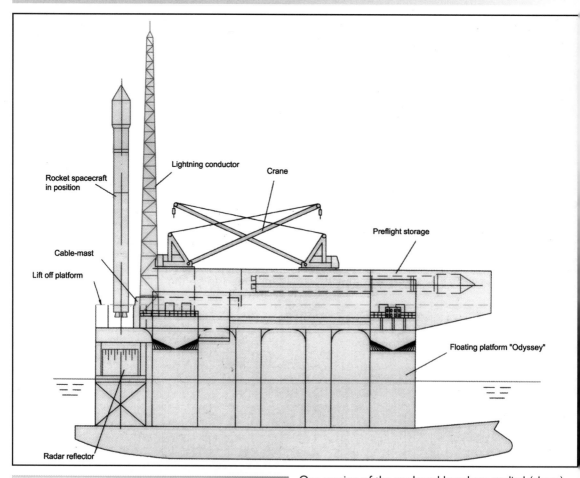

Rocket spacecraft in position

Lightning conductor

Crane

Preflight storage

Cable-mast

Lift off platform

Floating platform "Odyssey"

Radar reflector

One version of the sea-based launchers studied (above).

Testing of a sea-based missile launcher.

Dates of Milestones in Rocket-Space Technology Creation at OKB-1 – TsKBEM – NPO Energia

May 13, 1946

The decision of the Government to form a number of Research Institutes (NII's), Design Bureaus (KB's), test organizations, and plants to develop, manufacture and test long-range ballistic rockets (LRBR). S. P. Korolev was appointed as the chief designer of liquid propellant LRBR.

October 18, 1947

The first launch of an LRBR in the Soviet Union – based on the German A4 (V-2) rocket.

September 17, 1948

The first launch of a native LRBR R-1. The rocket almost reached the specified range, but experienced a large deviation from the planned flight path because of abnormal operation of the control system.

October 10, 1948

The first successful launch of a native LRBR R-1.

April 21, 1949

The first launch (of six) of a geophysical rocket, the R-1A. Experiments with rocket head separation were performed on this rocket. The rocket lifted two instrumentation containers to an altitude of 100 km, which then landed by parachutes.

September 21, 1949

Launch of the R-2E rocket, an experimental check of the new R-2 rocket system's serviceability.

1950

The R-1 rocket complex is put into service.

1949-1951

The R-2 rocket, with a separable head, is created, then the R-2 complex is put into service.

1951-1956

Geophysical rockets R-1B, R-1E, R-1D and others, and the R-2A are created and launched. Upper atmospheric and space research is continued.

March 15, 1953

The first R-5 strategic rocket is launched. A modification of the R-5 (R-5M, first launched on January 21, 1955) was fitted with a special explosive charge. Geophysical rockets R-5A (launches in 1958-1961), R-5V (launches in 1964-1975, among them launches within the Vertical program), and others are created based on R-5 rockets.

April 18, 1953

The first launch of an R-11 tactical missile.

September 16, 1955

The first submarine launch of an R-11FM missile.

May 15, 1957

The first launch of an R-7 intercontinental ballistic two-stage missile.

August 21, 1957

The successful launch of an R-7 intercontinental ballistic two-stage missile.

October 4, 1957

The launch of the first artificial Earth-orbiting satellite, a mass of 83.6 kg. It remained in orbit for more than 92 days. On January 4, 1958 the satellite entered the dense upper atmosphere and burned up.

November 3, 1957

The launch of the second artificial satellite, of 508 kg mass, with dog Laika on board.

May 15, 1958

The launch of the third artificial satellite, a mass of 1,327 kg, by an R-7-type rocket with improved performance characteristics.

January 2, 1959

The launch of the first interplanetary station Luna-1 (Mechta) by an R-7 three-stage rocket, with a rocket unit E used as the third stage.

September 12, 1959

The launch of the Luna-2 station which delivered a USSR pennant to the Moon's surface on September 14, 1959.

October 14, 1959

The launch of the Luna-3 station, which photographed the back side of the Moon.

May 15, 1960

The launch of an unmanned Vostok-type spacecraft (1 KP).

August 19, 1960

The Vostok spacecraft (with dogs Belka and Strelka on board) is put into orbit. The animals were the first to be recovered from satellite orbit.

February 12, 1961
The Four-stage rocket (R-7 + rocket units I and L) puts into orbit an unmanned interplanetary station (UIS) Venera-1 (IVA No 2).

April 12, 1961
The first manned spacecraft – Vostok – (3KA) with Yuri Alexeyevich Gagarin on board goes into orbit.

April 26, 1962
The launch of a Zenit satellite to photograph the Earth's surface.

August 11-12, 1962
The first group space flight, comprised of the Vostok-3 and Vostok-4 spacecraft.

November 1, 1962
The unmanned interplanetary station Mars-1 (2MV-4 No 4) is put into orbit by a four-stage rocket.

January 30, 1964
The Electron-1 and Electron-2 satellites are launched by a single rocket to investigate the Earth's radiation belts (Van Allen belts).

October 12, 1964
The Voskhod multi-man spacecraft is put into orbit (3KV) – the first multi-man space flight.

March 18, 1965
The Voskhod-2 (3KD) spacecraft goes into orbit. A.A.Leonov makes the first ever egress into space.

1961-1968
The R-9, RT-1 and RT-2 rocket complexes are created. R-9 and RT-2 complexes are added to the national armory.

April 23, 1965
The launch of the Molniya-1 active retransmitter to provide an experimental long-distance radio communication line.

November 16, 1965
The launch of the Venera-3 unmanned interplanetary station which delivered a pennant to the surface of Venus on March 1, 1966.

January 31, 1966
The Luna-9 unmanned interplanetary station performs a soft landing on the Moon and transmits TV images of the Moon's surface to Earth.

March 10, 1967
The first (Zond) spacecraft launch of the L1 (7K-L1) program.

April 23, 1967
The launch of a new spacecraft – Soyuz-1 – with V. M. Komarov on board.

October 30, 1967
Automatic docking of Soyuz-type spacecraft (Cosmos-186 - Cosmos-188).

January 15, 1969
Docking of the Soyuz-4 and Soyuz-5 manned spacecraft. Cosmonauts transfer from one spacecraft to the other through outer space. Creation of an experimental station of 12,924 kg mass.

1961-1974
Work carried out on the Moon program to create a modular multi-purpose launch vehicle, N1, and a lunar complex, L3. On February 21, 1969, complex N1-L3 flight tests began. The program was canceled because of breakdown of the schedule for the lunar complex creation, and after four (out of four) launch failures.

April 19, 1971
The launch of the Salyut orbital station, which stayed in orbit until October 11, 1971.

June 30, 1971
The Soyuz-11 spacecraft goes into orbit, and then docks with the Salyut orbital station. This marks the beginning of manned flight mode operation for the Salyut station (which lasted 22 days).

December 26, 1974
The launch of the Salyut-4 station. It remained in orbit until February 3, 1977. Two crews operated on board the station.

July 15, 1975
The Soyuz-19 spacecraft is launched, which then docks to the U.S. Apollo spacecraft on July 17, 1975. The first experimental flight of a space complex comprised of spacecraft from two countries (the Apollo-Soyuz program).

1976
Beginning of work on the Energia versatile space transportation system and the Buran orbiter.

September 29, 1977
The Salyut-6 station – a station of the second generation – with two docking units is put into orbit. It remained in orbit until July 29, 1982. 16 crews operated on board the station.

December 10, 1977
The Soyuz-26 spacecraft goes into orbit, then docks with the Salyut-6 orbital station. This begins the Salyut-6 manned operation mode.

January 20, 1978
The first Progress unmanned cargo transport spacecraft flight. The first delivery of cargoes to the station by the transport spacecraft.

March 2, 1978
The Soyuz-28 spacecraft, with the first international crew on board, goes into orbit and docks with the Salyut-6 orbital station.

December 16, 1979
The Soyuz T first unmanned flight. It docks with the Salyut-6 orbital station and the Salyut-6 / Soyuz T complex flight continues for more than 100 days.

June 5, 1980
The Soyuz T-2 manned spacecraft is launched and docks with the Salyut-6 orbital station.

April 19, 1982
The Salyut-7 station – a Salyut-6 station back-up – is put into orbit. It remained in orbit until February 7, 1991. Ten crews operated on board the station.

February 20, 1986
The core module of the Mir permanent manned complex is put into orbit. Manned operation mode began on March 15, 1986. Three special-purpose modules (Kvant astrophysics module, launched on March 31, 1987; Kvant-2 add-on module, launched on December 6, 1989; and Kristall technological module, launched on May 31, 1990), as well as a Progress M-type cargo spacecraft and Soyuz TM-type transport spacecraft (with the main crew and a visiting one) are docked to the core module.

March 13, 1986
The Soyuz T-15 spacecraft is launched and docks with the Mir complex on March 15, beginning of the complex manned operation mode. Soyuz T-15 performs an orbital transfer to the Salyut-7 station and back to Mir (May 5 - June 26) and delivers 400 kg of cargo from Salyut-7 to the Mir complex for further use.

May 21, 1986
Docking of the first Soyuz TM spacecraft (Soyuz TM-1), in unmanned mode, to the Mir complex.

February 6, 1987
The launch of the Soyuz TM-2 manned spacecraft which docks with the Mir orbital complex.

May 15, 1987
The launch of an Energia launch vehicle with the Skif-DM spacecraft on external suspension.

November 15, 1988
An Energia launch vehicle launch with the Buran orbiter attached in an unmanned mode.

August 23, 1989
The launch of a Progress M unmanned cargo transport spacecraft.

1990
Beginning of work on the Energia-M launch vehicle.

September 27, 1990
The launch of a Progress M-5 cargo spacecraft with a recovery ballistic capsule on board which delivers the onboard results to the ground. The landing was performed on November 28, 1990.

December 2, 1990
Soyuz TM-11 - Mir Expedition with an international crew including T. Akiyama (a Japanese journalist), the first commercial passenger to Mir. Akiyama made daily television broadcasts.

1992
The beginning of extensive international activities in joint space exploration programs.

March 17, 1992
Soyuz TM-14 - Joint flight with Germany.

1993
Activities in the Mir complex program continue. The 14th expedition began to operate on board the Mir complex from July 1.

February 3-11, 1994
STS-60 was the first flight of a cosmonaut aboard the US Shuttle. Sergei K. Krikalev as a mission specialist conducted joint science programs.

November 12-20, 1994
STS-74 was the first shuttle assembly flight to Mir, it carried a Russian-built docking module with two attached solar arrays.

May 1995
The Spektr ("Spectrum") module joined Mir in May 1995. The module was designed for scientific research, specifically Earth observation. The final module was the Spektr Remote Sensing Payload. It had instruments to study particles in low Earth orbit. This module was damaged in the collision with a supply ship and was closed up pending final repairs that were never finally completed.

June 27 - July 7, 1995
STS-71 Atlantis performs the first US Shuttle docking with Mir.

1995
The docking module was added to Mir during the second US Shuttle / Mir docking mission, STS-74, in late 1995.

March 22-31, 1996
STS-76 began the continuous U.S. stay on Mir. A single Spacehab module was aboard, demonstrating logistics capabilities.

April 1996
The Priroda ("Nature") module was launched in April 1996, completing the assembly of the Mir complex. This module carried Earth observing equipment as well as experiments.

August 17, 1996
This launch was the first of the Soyuz-U boosters with a crew aboard.

February 1997
During February, a fire occurred aboard Mir, offering new challenges and new information. The first spacewalk by a U.S. astronaut outside Mir wearing a Russian spacesuit was made.

June 25, 1997
The Progress M-34 spacecraft crashes into Spektr. The collision damaged one of the solar panels and also punctured the hull, depressurizing the module.

September 25 - October 6, 1997
Astronaut Scott Parazynski and Cosmonaut Vladimir Titov conducted a joint spacewalk.

November 20, 1998
The Zarya ISS module is launched by a Proton rocket for rendezvous with the US Unity module. The hatch between Unity and Zarya is opened for the first time on Dec 10, 1998.

February 20, 1999
Soyuz TM-29 docked with Mir on February 22. After accepting a double-length assignment, Russian cosmonaut Viktor Afanasyev set a new cumulative time in space record, but then, for the first time since September 1989, there were no humans in space.

July 12, 2000
The Zvezda ISS module is launched by Proton rocket and docks with the ISS Zarya module on Jul 26. The ISS now consists of three modules: Zvezda, Zarya and Unity.

October 31, 2000
Soyuz TM-31 spacecraft launched by Soyuz-U rocket carrying the crew of the first ISS Expedition and docks with the ISS Zvezda module on November 2, 2000.

February 20, 2001
The core module of the Mir space station celebrates its fifteenth anniversary in orbit.

March 18, 2001
"Rock", the first of a pair of direct broadcast digital radio satellites is launched from the Sea Launch platform in the Pacific Ocean by a Zenit rocket into geosynchronous transfer orbit.

March 23, 2001
The Mir space station is deorbited successfully. Fragments of the world's most successful space station hit in a remote area of the Pacific following fifteen years of unprecedented orbital research.

To receive more information
about conclusion of agreement
on services with use
of space/rocket technology,
address:

KSCE
141070 Kaliningrad Moscow region, Lenin street, 4a
Telephone: (095) 513-72-48. Fax: (095) 187-98-77

The contribution of RSCE
to Russian space technology

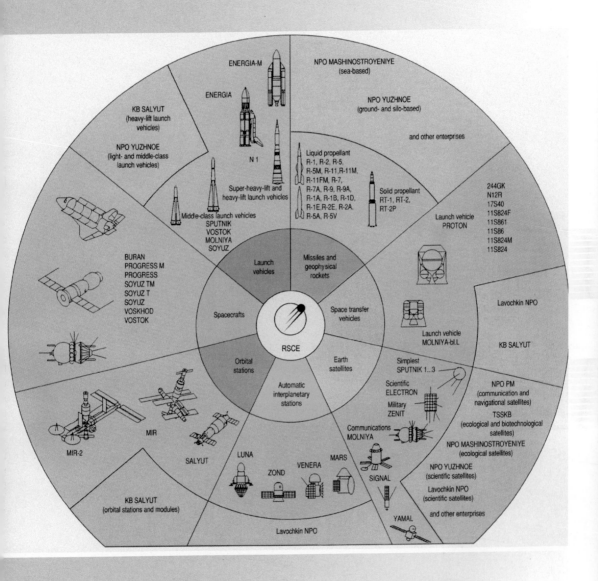

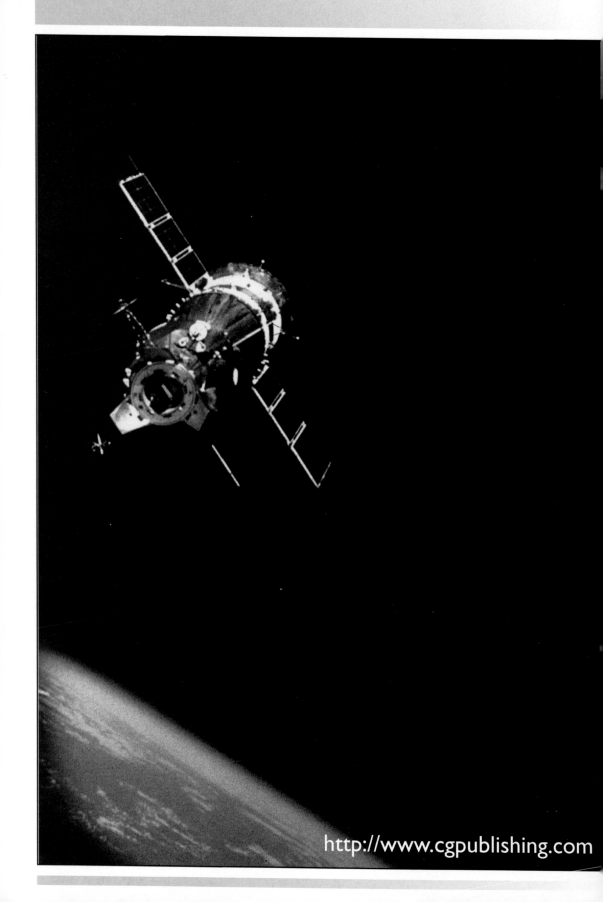